ALCHEMY AND AUTHORITY IN THE HOLY ROMAN EMPIRE

ALCHEMY AND AUTHORITY IN THE HOLY ROMAN EMPIRE

TARA NUMMEDAL

THE UNIVERSITY OF CHICAGO PRESS
CHICAGO AND LONDON

The University of Chicago Press, Chicago 60637
The University of Chicago Press, Ltd., London

Published 2007
Paperback edition 2019
Printed in the United States of America

28 27 26 25 24 23 22 21 20 19   1 2 3 4 5

ISBN-13: 978-0-226-60856-3 (cloth)
ISBN-13: 978-0-226-63972-7 (paper)
ISBN-13: 978-0-226-60857-0 (e-book)
DOI: https://doi.org/10.7208/chicago/9780226608570.001.0001

Library of Congress Cataloging-in-Publication Data

Nummedal, Tara
Alchemy and authority in the Holy Roman Empire / Tara Nummedal.
p. cm.
Includes bibliographical references and index.
ISBN 978-0-226-60856-3 (cloth : alk. paper) 1. Alchemy—History.
2. Alchemists—History—Sources. 3. Holy Roman Empire—History. I. Title.
QD13.N86 2007
540.1'12094—dc22

2007002521

∞ This paper meets the requirements of ANSI/NISO Z39.48-1992 (Permanence of Paper).

IN MEMORIAM
Lara J. Moore, 1971–2003

CONTENTS

FIGURES

ACKNOWLEDGMENTS

Paracelsus advised the student of nature to wander, to collect bits and pieces of wisdom from many different sources and locales. While I did not set out to follow his advice, this book is the product of many *Wanderjahren* among numerous universities, archives, and libraries. I owe an enormous debt to all of the colleagues I have met along the way.

I feel exceedingly fortunate to have worked with Paula Findlen at the University of California, Davis, and Stanford. She guided this project through its many different stages, always with the perfect balance of constructive criticism and encouragement. More broadly, she taught me how to ask questions about the past, and as a scholar, teacher, and mentor she continues to inspire by example. I also owe an immense debt to Pamela Smith, who not only first introduced me to the world of early modern alchemy in a course at Pomona College, but also convinced me that my interest was worth pursuing in graduate school. Her extraordinary generosity as a mentor, critic, and colleague has never flagged.

Numerous colleagues have offered rigorous and thoughtful critiques of this book along the way. While I was at Stanford, Brad Gregory, Katharine Park, and Nancy Kollmann read and offered challenging comments on numerous drafts of this project in its earliest stages, as did Gillian Weiss, Daniel Stolzenberg, Sara Pritchard, and Lara Moore, who is the one friend and colleague I will never be able to thank in person; both her smile and her intellect will be deeply missed. I was also enormously fortunate to spend a year in residence at the Max Planck Institute for the History of Science in Berlin while I scoured the archives of central Europe. I am enormously grateful to Lorraine Daston, Otto Sibum, and everyone involved in the seminar "Scientific Personae" at the institute in 1998–99; they provided the camaraderie and lively discussions that shaped the project from the beginning. My colleagues at Brown University and the University of Southern California have also been unfailingly supportive, despite my comings and goings. Deborah Cohen, Robert Self, Naoko Shibusawa, Mark Swislocki, Moshe Sluhovsky, Amy Remensnyder, and Jason Glenn have my particular gratitude for their critial engagement with my work, often despite its remoteness from their own. Oded Rabinovitch and Cryn Johannsen provided helpful research assistance. Finally, my deepest thanks to several "at large correspondents" who have provided delightful social and intellectual companionship over the years, especially Nicholas Dew, Hereward Tilton, Claudia Stein, Margaret Garber, Chris Carlsmith, and David Ciarlo. Alisha Rankin deserves special thanks for always sharing her expertise in the history of medicine, Saxony, and early modern German. So too does Janice Neri, who has read more drafts and chapters and indulged more anxieties than any friend should ever have to tolerate.

At the University of Chicago Press, Catherine Rice oversaw the early phases of this book, while Jennifer Howard heroically stepped in to see the project through with the utmost professionalism, efficiency, and good cheer. Special thanks go to Pete Beatty and Erik Carlson at Chicago for patiently fielding my many questions, and to Christopher Stanwood at the Chemical Heritage Foundation for his assistance with images. I am particularly indebted to the two anonymous readers for the press, both of whom offered important suggestions and critiques. I have also been fortunate to present parts of this project to many challenging audiences over the years in the United States, Canada, and Europe. I am deeply appreciative to all who engaged my ideas and offered feedback at seminars and conferences; their comments have shaped this book in immeasurable ways.

I could not have completed this project without the assistance of the librarians and archivists at the Niedersächsisches Staatsarchiv Wolfenbüttel, Sächsisches Haupstaatsarchiv Dresden, Württembergisches Hauptstaatsarchiv Stuttgart, Státní oblastní archiv Třeboň, Bayerisches Hauptstaatsarchiv München, Thüringisches Hauptstaatsarchiv Weimar, Huntington Library, and Wellcome Library. Generous financial support from the History Department at the University of California, Davis, the History Department and Social Science History Institute at Stanford University, the Center for German and European Studies at the University of California, Berkeley, the American Council of Learned Societies, the Deutsche Akademische Austauschdienst, the Mrs. Giles M. Whiting Foundation, Brown University, and the National Endowment for the Humanities allowed me the rare privilege of devoting my full attention to these alchemists. Any views, findings, conclusions, or recommendations expressed in this book do not necessarily reflect those of the National Endowment for the Humanities.

My extended family has heard about this book—and no doubt wondered about alchemy—for many years. I could not have completed this project without their support and understanding, especially that of Susan Glogovac and Earl Carlson. Finally, more than anyone, Seth Rockman supported and sustained me through this long project in ways I could never properly express.

NOTE ON EARLY MODERN WEIGHTS AND MEASURES

Weights, measures, and currency in the Holy Roman Empire were complicated by a lack of centralization. The following units of measure appear in this book:

COINS

Coin	Theoretical Weight	Theoretical Silver/Gold Content	Fineness[a]
Reichsthaler or thaler (silver)	29.23 g	25.98 g silver	889‰
Silbergulden (silver)	24.62 g	22.91 g silver	931‰
Mark (gold)	233.85 g	23⅔ carat gold	986‰
Ducat (gold)	3.49 g	3.44 g gold	979‰
Gold gulden or reinischer gold gulden (gold)	3.25 g	2.45 g gold	753‰

Sources: Helmut Kahnt and Bernd Knorr, Alte Maße, Münzen und Gewichte *(Mannheim: Bibliographisches Institut/Mayers Lexiconverlag, 1987); Jost Weyer,* Graf Wolfgang II. von Hohenlohe und die Alchemie: Alchemistische Studien in Schloß Weikersheim, 1587–1610, *ed. Historischen Verein für Württembergisch Franken, Stadtarchiv Schwäbisch Hall and Hohenlohe-Zentralarchiv Neuenstein, Forschungen aus Württembergisch Franken, vol. 39 (Sigmaringen: Jan Thorbecke Verlag, 1992), 409–14.*

Note: The Augsburger Reichsmünzordnung of 1556 declared the ducat to be the official unit of gold currency in the Holy Roman Empire. Its weight and fineness remained constant even after the dissolution of the empire in 1806. Although not quite as widely used as the ducat, the gold gulden (or rheinischer gold gulden) was widely used as a unit of currency in trade in the sixteenth century. The value of the thaler fluctuated according to time and place.

[a] *1‰* = 1 part per thousand.

WEIGHTS

Measure	Weight	Value
1 loth	14.62 g	
1 mark	233.85 g	16 loth
1 pfund	467.7 g	32 loth
1 zentner	46.77 kg	100 pfund

Source: Weyer, Graf Wolfgang II. von Hohenlohe und die Alchemie, *409–14.*

LIQUID MEASURES IN WÜRTTEMBERG

Measure	Volume
Eimer	293.9 l
Maß	1.84 l
Viertel	.45 l
Scheffel	177.2 l
Schoppen	.459 l

Source: Weyer, Graf Wolfgang II. von Hohenlohe und die Alchemie, *410.*

ABBREVIATIONS

NStA Wolfenbüttel	Niedersächsisches Staatsarchiv Wolfenbüttel
SHStA Dresden	Sächsisches Hauptstaatsarchiv Dresden
HStA Stuttgart	Württembergisches Hauptstaatsarchiv Stuttgart
SOA Třeboň RRA	Státní oblastní archiv Třeboň, Rožmberský roddiný archiv
BHStA Munich	Bayerisches Hauptstaatsarchiv München

INTRODUCTION

The Rise and Fall of Philipp Sömmering (ca. 1540–75)

Philipp Sömmering had much to offer Duke Julius of Braunschweig-Wolfenbüttel when he arrived at this northern German court in 1571. Like numerous other alchemists circulating among the cities and princely courts of the Holy Roman Empire, Sömmering claimed to know how to transmute base metals into silver and gold. The inventions and cures he had to offer, however, extended far beyond this core alchemical skill. Sömmering's ideas about improving productivity in the local mines certainly impressed Duke Julius, who took a keen interest in the metal and salt mines in his territories. The alchemist's design for a gun barrel that promised to shoot bullets absolutely straight was equally intriguing, as was his potential as a religious adviser. As a former pastor who claimed to be trained by Luther's close friend Melanchthon, the alchemist was well qualified to guide Duke Julius through the implementation of the Reformation in his formerly Catholic territories. Like other typical patrons of alchemy, Duke Julius appreciated the combination of intellectual, religious, medicinal, technical, and economic skills that alchemists had to offer and gladly welcomed Sömmering and his assistants to the Wolfenbüttel court.[1]

When Sömmering arrived in Wolfenbüttel, he negotiated a formal employment contract with his patron. The alchemist vowed to produce one loth (or 14.62 g) of a tincture for transmuting metals within a fixed period of time in exchange for the tidy sum of two thousand thaler. Sömmering promised to return Duke Julius's money should he fail. For his part, the duke gave Sömmering his advance as well as ten ells of English cloth, one hundred thaler, and a horse as tokens of his gratitude. To facilitate this alchemical work, Duke Julius provided the alchemist with both lodging and a laboratory in a former pharmacy. Sömmering also hired a number of assistants to help him complete his work.[2]

Philipp Sömmering sought to trade his alchemical knowledge for the security and prestige of Julius's patronage. He had come to Wolfenbüttel from Gotha, where he had experienced firsthand the chaos of the politico-religious warfare surrounding the Reformation.[3] From the very beginning of their stay in Wolfenbüttel, however, Sömmering and his assistants found that their refuge there was fragile. Above all, they faced strident opposition from Duke Julius's consort, Duchess Hedwig. Evidently some of Hedwig's ire was personally directed against Sömmering's fellow alchemist Anna Maria Zieglerin, who as a female alchemist seemed particularly threatening to the dominant woman at court. Mere months after her arrival in Wolfenbüttel in 1571, Zieglerin complained to Julius that "your beloved consort has hurled her blazing fury at us."[4] Sömmering shared this perception that the court had quickly become a hostile environment, lamenting the defamation of his good name in his own letter to the duke a few months later.[5] Sömmering and Zieglerin were certainly not imagining this ill will on the part of some members of Julius's court. When the duke's sister, the Margravess Katharina of Brandenburg, visited Wolfenbüttel in November of 1572, she too came to dislike and distrust the alchemists intensely. In a letter to Duchess Hedwig the following summer, Katharina told her sister-in-law how much it pained her to think that Sömmering and "that hag" (*losen Vettel*) Zieglerin "are around Your Dearest and Your young reign almost every day and have moved to Your Dearest's table so that Your Dearest must eat and drink with them." "God knows I mean this truly and from my heart," Katharina pledged, "I will not cease to ask God to enlighten Your Dearest's and my dear brother's heart and spirit and turn it away from these cursed, sinful, godless people. Wicked company can cause things to happen which otherwise would never happen; remember that a lord is judged by his servants and advisers."[6]

In the fall of 1573, Katharina's prayers seem to have been answered when Duke Julius's adviser Jobst Kettwig initiated a series of actions that would lead to the alchemists' downfall. Kettwig had obtained his prestigious position as *Kriegs- und Kammerrat* (military adviser and counselor) at Julius's court through his contact with one of Sömmering's assistants, but in November of 1573 Kettwig's criminal past caught up with him.[7] When the Danish crown issued a warrant for Kettwig's arrest as a *Straßenräuber und Landfriedensbrecher* (bandit and disturber of the peace),

Sömmering responded to the warrant and turned him in.[8] Imprisoned in nearby Braunschweig and furious with this turn of events, Kettwig took his revenge on the alchemist. In a fierce letter to the chief administrator in Wolfenbüttel, Kettwig claimed that Sömmering and his assistants were frauds. He alleged that they had presented themselves under false names and title, and that in fact they knew nothing of nature's secrets. According to Kettwig, in fact, the alchemists' entire pretense at court had simply been a ruse to take Julius's money.[9]

As the one who landed Kettwig in prison in the first place, Sömmering was an obvious target for Kettwig's wrath. As transparent as the motivations behind Kettwig's denunciation were, however, Sömmering realized that it could easily be his undoing. Hoping to quiet Kettwig by facilitating his escape, Sömmering made a mistake that literally turned out to be fatal. Days after Kettwig's disastrous denunciation had reached the court, Sömmering helped him break out of his Braunschweig prison and flee Duke Julius's territory.[10] Meanwhile, Julius received a letter from the elector of Brandenburg that further impugned the honesty of the court alchemists.[11] Julius finally locked up Philipp, Anna, and several of their assistants in the Wolfenbüttel palace prison on Whitsunday, 1574, presumably to the great satisfaction of Margravess Katharina and Duchess Hedwig.

Over the course of the yearlong interrogation and trial, a multitude of charges against Sömmering and his assistants emerged. After conducting preliminary interrogations of Sömmering and a few other witnesses and debating whether to use torture, the court moved on to its more serious investigations in July 1574. According to the allegations, the alchemists were not simply impostors who stole Duke Julius's money; their crimes included adultery, murder (of a courier), attempted poisoning (of Duchess Hedwig), sorcery (involving a variety of charms, including a love philter and a formula for becoming invisible), illicitly copying keys to Duke Julius's chambers, stealing state papers, and inventing an imaginary alchemical consultant, Count Carl von Oettingen. Nearly every day during the month of July 1574, court officials interrogated Sömmering and other witnesses in the case, both with and without torture. After a hiatus in late summer and fall, the court met again at the end of November to continue the interrogations. After Sömmering and three others eventually confessed under torture to most of the crimes, in December the court sent its opinion to legal consultants in Magdeburg, Brandenburg, and Wittenberg for a verdict. The alchemists were condemned to death.[12]

The executions took place in February of 1575. Sömmering and one of his assistants were dragged to the Mühlentor, their skin torn with red-hot, glowing tongs, and their bodies quartered alive and posted on the four roads into town.[13] Another of Sömmering's assistants was also dragged to the Mühlentor. His body was first broken on the wheel before being quartered and hung up at the crossroads. Like the rest, Anna Zieglerin's skin was also torn with hot tongs, and she was then strapped to an iron chair and burned to death. Sömmering and his colleagues' careers came to an extraordinarily spectacular and gruesome finale that winter, even by early modern standards of punishment. Their lives and work as alchemists were far less anomalous, however. In fact, there is no better introduction to the high-risk, high-reward game of alchemy in the early modern Holy Roman Empire.

Alchemy in Practice and the Problem of Fraud

Philipp Sömmering's career in Wolfenbüttel dramatically illustrates the fact that, in the sixteenth and seventeenth centuries, alchemists were not simply figments of the European literary or philosophical imagination, but very real purveyors of practical techniques, inventions, and cures. Gathering their skills from a variety of sources, a surprising array of people came to call themselves alchemists in early modern central Europe. In princely laboratories, mining towns, and urban centers, alchemists entered into contracts to make precious metals, advised on mining projects, experimented with chemical medicines, and occasionally made gemstones or pearls as well. Others profited from alchemy by selling specific recipes and processes to wealthy patrons or fellow alchemical enthusiasts eager to explore and profit from nature's secrets. Although some alchemists practiced their art successfully and many even made careers of it, others, like Philipp Sömmering, found themselves facing the accusation that they were *Betrüger:* frauds or impostors.

This book follows Sömmering and his contemporaries as they struggled to define what it meant to practice alchemy in the early modern Holy Roman Empire. The dramatic burst of interest in alchemy in this period was accompanied by a great deal of disagreement about some of the most fundamental facets of alchemical practice. What was the best way to gain alchemical expertise, and what kinds of things could alchemists actually produce? Was alchemy simply a way to create medicines and precious metals, or must practitioners combine that work with a more sophisticated intellectual or even spiritual project? Should they sell their knowledge to patrons or other practitioners? Debates about alchemical ideas and practices were hardly new in the sixteenth century, of course; from the time that alchemy appeared in the Latin West in the twelfth century (and certainly for much longer in Arabic- and Greek-speaking cultures), alchemists and commentators had deliberated over the goals, ideas, uses, and practices associated with this ancient art. Early modern Europeans offered new answers to these age-old questions, embedding alchemy in the particular structures and problems of sixteenth- and seventeenth-century Europe. What is especially striking about the early modern debates, however, is that they also posed a new set of questions about the social dimension of alchemical practice, making the figure of the *alchemist* inseparable from the ideas and practices of alchemy. Could anyone become an alchemist? Was the alchemist merely an abstract symbol of folly or deception, or was he (or perhaps even she?) the type of person one might meet in the marketplace? Was the alchemist just another kind of skilled artisan who happened to specialize in metals, or did this new character more closely resemble something else—a prophet, perhaps, a scholar, courtier, expert consultant, or even a bureaucrat? Did the alchemist bear any recognizable social characteristics at all?

Early modern central Europeans explored these questions through both word and action, in private letters and published treatises, literary satires, paintings, courtrooms, assaying houses, and legal contracts. Scholarly alchemists wrote treatises asserting their identity as alchemists and laying out the fundamental ideas, practices, and goals of their art; these texts found their way into print in increasing numbers in the sixteenth century, as print culture and alchemical culture developed hand in hand. Alchemists appeared in other kinds of printed texts as well, as the subjects of humanist satires and images, but the world of print was by no means the only site for consideration of alchemy and its practitioners. Patrons routinely had to make judgments about alchemy when practitioners petitioned for positions in courtly laboratories and performed demonstrations of their art. When alchemists like Sömmering were turned over to the courts under charges of fraud, legal authorities took up questions about alchemy and its practitioners from within the parameters of jurisprudence. Finally, early modern Europeans confronted questions about alchemists (albeit in ways that are often lost to us) whenever they met one, perhaps at the tavern, the market, or the princely court. Printed debates about alchemists and their activities, in other words, emerged alongside and in dialogue with the real men and women who were putting alchemy into practice. Alchemists' activities shaped the contours of the debate, which responded to their successes and failures; in turn, these debates shaped alchemists' lives and work by creating new expectations and standards of legitimacy.

Early modern discussions about the nature and limits of alchemy, then, were never divorced from the figure of the alchemist, who personified the debate and anchored it to the society and culture of early modern Europe. Strikingly, central European efforts to define this elusive figure were often articulated in the negative, by creating a kind of alchemical evil twin: the fraudulent alchemist, or, in German, *Betrüger.* For the modern observer, the idea of a fraudulent alchemist might seem curious, even redundant. Aren't all alchemists frauds? Because we now believe that one of early modern alchemy's central claims—the transmutation of base metals into gold—is impossible, we tend to assume that anyone who professed the ability do so in the past was either a cheat or a fool. To Sömmering's contemporaries in the Holy Roman Empire, however, the link between alchemy and fraud was not yet naturalized, cemented by modern categories of truth and falsehood. To be sure, there was little doubt in the sixteenth century that *Betrüger* existed; from the most scholarly alchemists to the casual tinkerer, from alchemy's most eloquent promoters to its fiercest critics, nearly all could agree that fraudulent alchemists were real and that they were a threat. However, few would have gone so far as to conflate those false alchemists with all alchemists. Rather, the *Betrüger* were thought to be a problematic subcategory of alchemists, akin perhaps to the false beggars who lurked among the honest poor in early modern cities. While early modern central Europeans acknowledged the existence of *Betrüger,* they not only made careful distinctions between those false alchemists and the true practitioners of the art; they devoted enormous energy to maintaining the border between these categories.

By personifying what the alchemist was *not,* therefore, the figure of the fraudulent alchemist—the *Betrüger*—became a central locus for working out what alchemy *was* in the Holy Roman Empire. The notion of the *Betrüger* also resonated beyond those immediately involved with alchemy, because it configured the alchemist as moral exemplar, a symbol of deceit and folly. Humanists and artists took up this theme, turning the internal debates among alchemists and their patrons to entirely different cultural ends. The notion of fraud, then, became not only a way for alchemists to discuss their own art, but also a focal point for a much broader discussion about alchemy's cultural meaning. For this reason, this book takes fraud as a point of entry into early modern alchemy, exploring the many valences of this concept and how they functioned to define and shape alchemical practice and its cultural meaning in the early modern Holy Roman Empire. Put differently, this is a social history of alchemy in central Europe, and a cultural history of why it proved to be so contentious.

This book begins, with some very basic questions. What kinds of people actually practiced alchemy in early modern central Europe, and where did they acquire their skills? Why were wealthy patrons interested in alchemical practices, and how did they decide whom to hire? What were real-life laboratories like? Finally, what exactly counted as alchemy in the eyes of early modern central Europeans? In order to reconstruct the material and social practice of alchemy, this book draws upon the numerous letters, laboratory reports, and supply orders connected with alchemists' employment across the Holy Roman Empire. The abundance of such materials in the archives of central Europe is itself a testimony to the important presence that alchemy once had there. Alchemists wrote letters proposing processes to potential patrons; successful proposals resulted in contract negotiations over salary, expectations, and materials. Functioning laboratories left multiple archival (not to mention archaeological) traces as well: work contracts and architectural plans for constructing or converting buildings, employment contracts for staffing them, and inventories taken upon a patron's death. Unsuccessful alchemical projects were even more likely to generate paperwork. Disgruntled patrons who occasionally prosecuted failed alchemists for fraud also collected testimony, interrogation transcripts, letters, and official reports as evidence. Collectively, these documents reveal a great deal about what it meant to be an alchemist in early modern central Europe. Like the documents from Duke Julius's court in Wolfenbüttel, these sources speak to who alchemists were, how they gained their knowledge and found employment, what projects they pursued, where they worked, and what happened when they failed.

Court administrators amassed many of these documents in the context of trials for alchemical fraud. Most historians will be quite familiar with these kinds of criminal records. Scholars of early modern Europe in particular have used them frequently to access the lives of individuals who otherwise did not leave a trace in archives and libraries.[14] Historians of science, however, have used trial records much less frequently, particularly those from secular courts.[15] These criminal records—particularly the interrogation records—are, of course, notoriously difficult to work with. The voices of accused criminals have come to us in highly mediated form, and the power dynamic involved in confessions recorded under the threat or use of torture makes them particularly problematic. As Natalie Zemon Davis has noted, however, the tales people told, even under the pressure of life and death, "can still be analyzed in terms of the life and values of the person saving his neck by a story."[16] Criminal records may never tell us what "really" happened, but the fact that witnesses' stories had to be at least plausible makes them important sources for contemporary expectations. Philipp Sömmering and his accomplices, for instance, credited their fellow alchemist Anna Zieglerin as the source of most of their ideas. Her role may have been this central; alternatively, she may also have been merely a scapegoat. In either case, the fact that everyone involved in her case found it convincing that a woman could even *be* an alchemist, let alone a source of alchemical ideas, demonstrates that a female alchemist was at least within the limits of the conceivable; this, in itself, is an important insight. Criminal records yield a great deal of less problematic information as well. In addition to biographical details about just who practiced alchemy, for example, trial dossiers contain patronage appeals and laboratory reports, written to princes long before legal proceedings began. Such documents often provide detailed information about the day-to-day activities of working alchemists. More broadly, they provide a sense of how practitioners understood alchemy and responded to its promise as well as its occasional betrayal.

In attempting a broad history of alchemical practice in early modern central Europe, this book both relies on and departs from previous work on alchemy. My work is deeply indebted to historians and historians of science who have demonstrated alchemy's important place in the intellectual landscape of early modern Europe. Whereas an earlier generation of scholars saw alchemy either as irrelevant or as a hindrance to the development of modern science, more recent work has demonstrated in various ways that alchemy was in fact central to what is still most often known as the Scientific Revolution. Recent scholarship has placed alchemical ideas and practices as the driving force behind the emergence of laboratories, debates about the power of human technology and the boundary between art and nature, matter theory, and even specific ideas like gravity. Precisely what role alchemy played in the emergence of modern science is still a matter of vital debate, but few would find it possible to talk about the Scientific Revolution without mentioning alchemy, if only because so many of the major figures in seventeenth-century science, including Isaac Newton and Robert Boyle, were fascinated by alchemy and deeply engaged with it.[17]

This book also draws on a wealth of scholarship on princely courts as sites for the production of natural knowledge in general, and alchemy in particular. The nobility was often deeply interested in collecting, studying, and manipulating nature in early modern Europe. Princes added spaces in their palaces for collections, libraries, and laboratories and devoted their resources to the work of all kinds of people with expertise about nature, from natural philosophers to engineers, physicians, cosmographers, artists, and surveyors. This princely interest in nature was motivated by any number of factors, ranging from the intellectual engagement and curiosity of what Bruce Moran has called "prince-practitioners" to a recognition that nature offered both practical expertise useful in state building and commerce and ideological justifications of power.[18]

Alchemy was ideally situated to appeal on these multiple levels because it was such a versatile art; working at the intersection of faith, commerce, and knowledge, alchemists like Sömmering addressed simultaneously many different facets of their world. In studies of noble patrons who devoted their intellectual and financial energies toward alchemical pursuits, historians such as R. J. W. Evans, Bruce Moran, and Pamela Smith have found that alchemy had a particular ideological resonance with early modern rulers. This recent work on the Holy Roman Empire has shown that princes such as Holy Roman Emperor Rudolf II (r. 1576–1608) or Landgrave Moritz "the Learned" of Hessen-Kassel (r. 1592–1627) embraced alchemical philosophy and natural magic in part because these philosophies reinforced other princely agendas. For these princes, alchemy and the study of nature's secrets could be many things: an expression of religious tolerance in the face of the tensions of Reformation Europe; a type of symbolic political propaganda that equated the control of the natural world with control of the human world; or a familiar language in which to couch more novel proposals, such as capitalist ventures. In contrast to a traditional view that saw princely patronage of alchemy as either irrelevant or an attempt to withdraw from an increasingly difficult political world into arcane mysticism, this work collectively argues that alchemy was a part of an active engagement with that world, part of an official court philosophy.[19]

Intellectual historians and historians of science, therefore, have deepened our understanding of the relationship between alchemy and the Scientific Revolution as well as the political and religious context surrounding its patronage. We still know surprisingly little, however, about the lives and labor of the vast majority of alchemical practitioners themselves and how they fitted into the social and cultural fabric of early modern Europe. Recent work on well-known figures such as Michael Maier, Johann Joachim Becher, George Starkey, Robert Boyle, and Isaac Newton, all of whom were deeply engaged with alchemy, has been an essential first step. All of these men were prolific authors with a range of interests, which has made it possible to examine the relationship between alchemy and other fields like political economy, natural philosophy, and biblical chronology. Moreover, by directing scholarly attention to the alchemical ideas and practices of such prominent figures as Isaac Newton, historians have shown that alchemy was a fixture in the mainstream intellectual landscape of early modern Europe, rather than a quirky side project or a fringe philosophy. In a nutshell, if Isaac Newton took alchemy seriously, so must we.[20]

There is also a disadvantage to putting figures such as Newton and Boyle at the center of a history of alchemy, however, in that it can create a misleading impression of the typical early modern alchemist. For an average German *Bürger* in 1620, the appellation "alchemist" would probably conjure up someone like Philipp Sömmering, not an Isaac Newton. Few alchemists were erudite, Latin-speaking scholars who were able to see their ideas into print, nor were they "important" in the traditional sense of contributing new alchemical theories or practices. We might think of these more workaday practitioners as entrepreneurial alchemists, in the sense that they practiced alchemy not solely for their own edification, but also for the social and material benefits it offered. Nevertheless, if the majority of alchemists lacked the scholarly distinction of someone like Isaac Newton, their practices were an important part of the engagement with alchemy in the Holy Roman Empire. Court ladies and gentlemen, apothecaries, pastors, and urban consumers of vernacular books of secrets captured the interest of wealthy patrons with their alchemical skill and knowledge, traded recipes and techniques with fellow practitioners, and staffed the growing number of laboratories that dotted central Europe. Numbers alone make these entrepreneurial alchemists a significant enough presence in the cities and courts of the Holy Roman Empire that they should take center stage in any history of alchemy. In one sense, then, this book offers a new perspective by writing the history of alchemy "from below," placing these more numerous ordinary alchemists, even the failed *Betrüger*, at the center of the narrative, rather than the better-known scholarly alchemists or their princely patrons.

As one might expect, early modern alchemy looks quite different from this point of view. It is a central contention of this book, however, that these entrepreneurial alchemists did more than simply add to a centuries-old alchemical tradition; they fundamentally transformed it. Together with their better-known contemporaries, these unknown alchemists created an incredibly rich and varied alchemical practice that threatened to burst the bounds of alchemical tradition. The collective presence of *all* of these types of alchemists—some of whom were quite successful, and others of whom were spectacular failures—forced a discussion about alchemy in early modern central Europe that would have lasting consequences. The resulting debate about true and false alchemy, moreover, was never merely rhetorical, but developed in dialogue with alchemical practices and careers of all kinds of alchemists, natural philosophers, and entrepreneurs alike—even supposed *Betrüger* like Philipp Sömmering—through the sixteenth and seventeenth centuries. For this reason, even if they did not make clear contributions to Europe's intellectual history, entrepreneurial alchemists are crucial not only to any understanding of alchemical practice and its cultural positioning in early modern society and culture, but also to any understanding of alchemy's relationship to state formation, court culture, natural knowledge, and social identities in early modern central Europe.

Alchemy was not alone in attracting this diversity of ideas and practitioners. This project also draws on a great deal of recent work by historians of science who have uncovered the sheer variety of what has come to be called natural knowledge in early modern Europe. This broad term encompasses not only what early modern Europeans called natural philosophy, the traditional focal point of the Scientific Revolution, but also the kind of knowledge and skill possessed by artisans, naturalists, collectors, instrument makers, navigators, gardeners, medical practitioners, and even artists and theologians, all of whom were actively engaged with nature in countless ways. As older systems that governed the pursuit of natural knowledge began to break down in the sixteenth century, these groups asserted numerous ideas about what might replace ancient and medieval knowledge. In every field, this diversity was fruitful and stimulating, but it also produced conflict. Battles between different kinds of practitioners—such as "charlatans" and physicians, navigators and cosmographers, or specimen peddlers and naturalists—mushroomed in medicine, navigation, and natural history, just as they did among alchemists. The debate about alchemical fraud, then, also goes to the heart of much broader issues about authority and knowledge in early modern Europe, offering new insight into the ways in which older standards of intellectual legitimacy coexisted and more often competed with more novel measures of authority. These measures drew on new commercial, political, and social contexts for natural knowledge, firmly resituating it in a uniquely early modern context.[21]

As historians of science have begun to follow natural knowledge beyond universities and academies and into princely courts, artisanal workshops, households, museums, marketplaces, and ships bound for the Americas, they have brought the history of science into much closer dialogue with the broader history of Europe. This book aims to contribute to this process by engaging issues of central concern to social and cultural historians of early modern Europe: the organization of work, social mobility, court culture, and crime. Too easily relegated to the twin stereotypes of adept or fraud, alchemists typically are assumed to have lived only at the margins of early modern society, either as criminal con men or otherworldly philosophers. Somehow they are assumed to have existed outside the normal social structures, work patterns, and practical concerns of everyday early modern Europe. Approached from the perspective of early modern social and cultural history, however, people like Philipp Sömmering provide an opportunity to step beyond stereotypes to explore the diverse lives and labor of actual alchemists. We find alchemical practitioners at work in laboratories, constantly interrupted by deliveries of food, visits from patrons, and uncooperative coworkers, not in cells hermetically sealed from the outside world. For many in central Europe, alchemy potentially was even a very profitable career, only one of many entrepreneurial ways to climb the social ladder and cross boundaries—both social and geographic—in the early modern period. Alchemists offer an example of a particularly early modern figure, a new, recognizable social type whose status depended not on birth, but on skill and talent; in this sense, alchemists participated in the practices of self-construction—ranging from self-fashioning to dissimulation—that historians have identified as particularly characteristic of early modern Europe.[22] In other words, alchemists operated according to many of the same rules as the rest of early modern society. Like any courtier, they sought princely favor; like other skilled artisans, they sought work contracts; and like so many others, they sometimes fell afoul of the law, ending their lives on the gallows. By looking at the many representations of alchemists in art, literature, and treatises in conjunction with the day-to-day lives of practitioners, this book seeks to reintegrate them into the social and cultural landscape of early modern Europe just as previous historians have integrated them into its intellectual landscape.

Finally, a focus on entrepreneurial alchemical practices highlights the important role that alchemy played in the economy of central Europe, for alchemists' work placed them firmly in the midst of the worldly negotiations of contracts, commerce, and state formation. Rather than operating outside a burgeoning commercial economy, the growing pool of people who claimed to be alchemists in the sixteenth century took advantage of it. They peddled practical books, secrets, techniques, and labor in a vibrant market for alchemical goods and services. Princes and financial elites were the most visible consumers of this entrepreneurial alchemy because they had the resources to dedicate to large alchemical projects. They also had particular reasons to be interested in alchemy in the late sixteenth century. Nearly constant warfare, Renaissance building projects, and courtly splendor drained princely coffers. At the same time, the rich central European mines, long a fundamental part of the economy of the Holy Roman Empire, began to stagnate while silver poured into Spanish coffers from the Americas. By 1600, the golden age of central European mining had come to a close and powerful investors (such as Duke Julius in Wolfenbüttel) were looking for new ways to sustain and develop the mining enterprises in their territories. Investing in alchemists was one solution. These efforts to use alchemy to address problems of mining and territorial finance suggest a more prominent place for alchemy in European economic history, not just as a metaphor, but also as a practice. This notion is certainly easy to dismiss. While we might be able to appreciate alchemy as an abstract philosophy, it is much more difficult for twenty-first-century observers to imagine it as a practical technology. Early modern patrons and practitioners viewed it as precisely such a technology, however, which could offer a solution to early modern economic concerns.[23]

In addressing these questions of social mobility and economy, this book links the alchemical study of nature to questions central to early modern European historiography as a whole. The particular geographic focus of my research, however, is the Holy Roman Empire. The importance of mining in central Europe made alchemy seem especially promising there. A sizable pool of workers with metallurgical expertise could potentially parlay that skill into alchemical careers, while the long-standing involvement of princes and banking families in mining enterprises in this region paved the way for elite support of entrepreneurial alchemical projects. Because central Europe was particularly fertile ground for alchemical pursuits in the early modern period, it offers an excellent microcosm of early modern alchemical practice. The sheer number of princes in the empire created a large pool of potential patrons, while the diversity of courts ensured that several varieties of alchemy could find support.

Of course, alchemy had different resonances in each princely court and city, as studies of patronage at single courts have shown. Some patrons were specifically interested in alchemical medicine, for instance, while patrons and practitioners in the mining regions of central and eastern Europe often were especially interested in the metallurgical facets of alchemical practice. This book focuses on archival research from several courts in particular—Dresden and Třeboň (known in German as Wittingau) in the east, Stuttgart and Munich in the south, Wolfenbüttel in the north; I also draw on existing studies of alchemy in Weikersheim, Kassell and Prague. While I hope that this project has remained sensitive to such local variations, the resources from these courts make possible a broad geographic scope, a survey of alchemical practice in the Holy Roman Empire as a whole (including Bohemia, which is often left out of studies of the empire).

Noble courts occupy a central place in this study because they were the most prominent sites of alchemical practice in early modern central Europe and their archives bear rich documentation of that practice. The sources collected at princely courts, however, often reveal a peripheral world as well. In letters and interrogations, practitioners frequently recounted details of their lives before coming to court, relating how they gained their expertise, why they chose alchemy as a career, and what they thought they could accomplish with it. These hints suggest that, while noble courts certainly served as the focus of alchemical practice, they could do so only by drawing upon a much more diffuse tradition of knowledge and practice. The focus of this book mirrors this configuration, placing the court at the center of the study while still acknowledging that alchemical practice was by no means limited to it.

Given the variety of practitioners and contexts that make up this book, it is difficult to point to a single set of ideas, practices, or even linguistic usages that clearly identify something as *alchemical,* as opposed to, say, medical, metallurgical, or natural-philosophical. As we enter the world of early modern alchemy, it is essential that we preserve this messiness, rather than trying to oversimplify alchemy or, worse, anachronistically to privilege one kind of alchemy over others based on our modern categories. In the sixteenth and seventeenth centuries, there was as yet no consensus about how to define alchemy. Anyone with an interest could approach the art from a variety of directions: through books, through informal conversations with other practitioners, or from related fields such as metallurgy, medicine, or natural philosophy more generally. Not surprisingly, this diversity meant that what practitioners understood to be "alchemy" included a variety of practices, skills, and knowledge bases. Today, we think of many of these as something else—pharmacy, perhaps, or metallurgy. In the early modern context, however, the only appropriate definition of alchemy is one that encompasses all of these activities and ideas. Although practitioners began to take sides on this issue as the sixteenth century came to a close, the battle to determine what alchemy would be was still far from won. For this reason, it makes more sense to think in terms of early modern *alchemies,* directing our eye toward the conflicts, contradictions, and multiple layers of meaning that characterized early modern European alchemical practices.[24]

Still, we can distinguish several different emphases within early modern alchemy, as long as we recognize that these categories were by no means fixed or exclusive; a single alchemist might pursue all of them, or only one, all the while claiming to be an alchemist. Wolf-Dieter Müller-Jahncke and Julian Paulus have suggested an evenhanded terminology that avoids the thorny issue of the qualities of the practitioners themselves (i.e., their education or intentions). Instead, Müller-Jahncke and Paulus stress the diversity within the alchemical tradition itself, a legacy of alchemy's many historical uses and the differences among its innumerable authorities. They distinguish between alchemy used primarily to produce medicines, or *alchemia medica,* and that aimed foremost at the production of precious metals, or *alchemia transmutatoria.* To *alchemia technica* Müller-Jahncke and Paulus ascribe "recipes oriented toward praxis" such as metallurgy and dye and ink manufacture, while *alchemia mystica* encompasses the more "mystical-christological" varieties of alchemy that sought above all understanding of God through his work in nature. These terms, even if they were not the actors' own, capture very well the different directions in which the alchemical tradition pulled its early modern practitioners, while quite rightly including them all within the broad category of alchemy.

Each of these varieties of alchemy had legitimacy in early modern central Europe in terms of the prestige and support associated with elite patronage; therefore, I have tried to avoid the claim that one sort of alchemy (the scholarly variety, for instance) was more genuine than others. Nevertheless, practitioners themselves often perceived contradictions among the various strains in the alchemical tradition and criticized their colleagues for privileging the wrong aspect of the alchemical arts. These debates often centered on the goal of alchemy or the identity of its practitioner more than the actual practices themselves. It would have been difficult to distinguish between the day-to-day alchemical operations of a scholarly alchemist and those of a more practical "goldmaker"; both worked in the laboratory, even if the goals of their alchemical work were different, if not contradictory. Disagreements about alchemy's goals—whether it was to aid in healing, a reform of knowledge, the production of things, or an understanding of God—exposed rifts in the Holy Roman Empire's community of alchemical practitioners. Such moments of conflict and definition form the centerpiece of this book. As some alchemists turned to print to set out precisely who was and who was not an alchemist, and as princes set out the terms of successful alchemical practice in contracts and trials for fraud, they exposed their own understandings of the art. Taken as a whole, these moments of definition reveal a stunning diversity of alchemies and alchemists.

One final word: people actually did alchemy in early modern Europe. When they called themselves alchemists and others found this believable, I take them at their word and call them alchemists as well, whether they claimed to be able to transmute metals, access divine secrets, assay metals, or produce fantastic medicines. Whatever we think of alchemy today, it is essential to remember that many people accepted the basic principles of alchemy in early modern Europe, even the transmutation of metals, and could point to religious and natural philosophical justifications for their belief. Successful alchemical works were rare and wondrous indeed, but not impossible. Thus, when we read in the newsletter of the Fugger banking family, for example, that the alchemist Marco Bragadino "changed a pound of quicksilver into gold some days ago" in Venice, we must accept that, in the eyes of early modern Europeans, he did indeed transmute metals.[25]

CHAPTER ONE

Assembling Expertise

How did one become an alchemist in early modern Europe? One might imagine joining the ranks of the adepts to be quite difficult. After all, as one mid-sixteenth-century alchemist put it, "alchemy is well hidden because the old masters who found the art did not want to teach [it to] either their children or their friends; therefore, he who finds this art is lucky, because it is not easily found."[1] Alchemical texts repeatedly described alchemy as a buried treasure, "the great gemstone and most noble pearl,"[2] and authors warned that the alchemist "should be secretive and silent and should reveal his secret to no one."[3] Alchemy, it seemed, was shrouded in secrecy, available only to the lucky treasure hunter or to one worthy to receive a revelation. Complicating matters further, alchemy also fell between the two institutions that traditionally oversaw the transfer of knowledge and expertise in early modern Europe: guilds and universities. Towns did not organize alchemists into guilds as they did members of other professions, such as surgeons, butchers, or tailors, suggesting that alchemy was not quite a craft trade. If interested and hopeful adepts could not sign up for guild apprenticeships, neither could they enroll at a university to access alchemical secrets. Perhaps suspicious of alchemy's status as a manual art, or perhaps more deeply skeptical about the philosophical status of alchemists' claims to surpass nature, the medieval founders of European universities had not integrated alchemy formally into the scholastic curriculum.[4] What was an aspiring alchemist to do?

Despite this lack of obvious venues for alchemical training in the sixteenth century, the number of men and even women who claimed to be practitioners suggests that the secretive and decentralized art was much more accessible than it would appear. Commenting in 1617 on the ease with which individuals turned themselves into alchemists, one rather cynical observer noted: "Look, and you will find [the alchemist's] primary transmutation to be of himself: a goldsmith becomes a goldmaker, an apothecary a chemical physician, a barber a Paracelsian, one who wastes his own patrimony turns into one who spends the gold and goods of others."[5] Indeed, the range of people engaged in the alchemical arts in central Europe was astounding. Not only natural philosophers and Paracelsian physicians, but also court ladies, pastors, apothecaries, Jews, and even "a maiden from Frankfort" studied alchemy and plied their art, if only to varying degrees of success.[6] Hardly the preserve of a few chosen adepts alone, alchemy was clearly accessible to a large section of the populace in the sixteenth century.

Where, then, did all of these alchemists acquire their skills and knowledge? This chapter explores the many different resources available to aspiring alchemists in the sixteenth-century Holy Roman Empire. As we might expect, some practitioners gained alchemical expertise by reading all or part of the vast corpus of treatises on the subject. Books were not the only resource, however. Overlapping fields such as metallurgy and medicine provided an invaluable source for essential skills and tools, as did the expertise of fellow practitioners of the alchemical arts. Although some alchemists argued that certain of these paths to alchemical wisdom were more legitimate than others, practitioners lacked the institutional authority to enforce those norms. Precisely because there was no alchemical guild or college to regulate practice and licensing, or an alchemy faculty at the universities to prescribe an authoritative curriculum, no single method of alchemical training prevailed. This lack of consensus opened the door to an unusually varied group of aspirants drawn from many segments of society; it also increasingly created battles over alchemical authority as the sixteenth century neared its end.

Alchemical Books and Their Renaissance Readers

In August 1573, when Philipp Sömmering advised Duke Julius how to pursue his interest in alchemy, the alchemist cautioned, "Such a philosophical secret should be learned and acquired not from the fraudulent processes of vagabonds [*landstreicher betrüglichen processen*], but rather from the most trustworthy books of the philosophers."[7] Although some natural philosophers, physicians, and other medical practitioners were increasingly looking beyond the written word and observing nature directly in the sixteenth century, texts were still a primary locus of natural knowledge. Immersed in this world, Sömmering thus counseled his patron that "the first step in coming to the proper understanding of the secret [of alchemy] is to get good, first-rate books."[8] But which books? By midcentury, literate students of alchemy were faced with a bewildering variety of texts. Manuscripts circulated widely, and modern print editions of ancient alchemical writings joined editions of medieval Latin and translated Arabic authors in the booksellers' shops. At the same time, new genres and texts flooded the market as practical *Kunstbüchlein* (or skills books) and the works of Paracelsus and his followers appeared in the bookstalls. Where should the would-be alchemist begin to read?

In order to guide his patron through this voluminous alchemical corpus, Sömmering provided Julius with a reading list that highlighted the importance of medieval Latin texts. Sömmering urged his patron to begin with the Catalan physician, religious reformer, and diplomat Arnald of Villanova (ca. 1240–1311), in whose name alchemical texts such as the *Rosarium philosophorum* appeared in the fourteenth century, because he offered a good foundation in both alchemical theory and practice.[9] According to Sömmering, the medieval physician explained the generation of metals and minerals in the earth and "describes splendidly" the basic components of all metals, mercury and sulfur; Arnald (or rather pseudo-Arnald, for it is unlikely that he wrote the texts that appeared in his name) also offered practical explanations of the "philosophical fire" and how to multiply metals by means of art (*künstlich*). "After that, read Roger Bacon [ca. 1215–after 1292] and Bernhard [of Treviso, fl. 1378]," Sömmering continued; "thus the philosophical fire will be thoroughly explained." Although Sömmering promised that Thomas Aquinas (ca. 1225–74) would "bring great understanding," he especially praised Bernhard as a font of knowledge about mercury, confessing that Bernhard's description of the "treasure of the secret," presumably the process for making the philosophers' stone, was his favorite book.[10] Finally, in addition to several other books that are now difficult to identify, Sömmering also included in his list an unpublished vernacular treatise by the fifteenth-century Bamberger Cathedral vicar and notary Johann Sternhals.[11] This treatise, entitled *Ritter Krieg* (War of the knights), was a "philosophical poem" in which gold (*Sol*) and iron (*Mars*) each argued their virtues (and slandered each other) in a court, with quicksilver (*Merkurio*) presiding as judge.[12]

Although Sömmering's list is perhaps an idiosyncratic slice of the alchemical corpus available to sixteenth-century readers, it nonetheless highlights the importance of a medieval Latin alchemical textual tradition that extended back to the late twelfth century. Alchemists, like Renaissance natural philosophers in general, had by no means embraced the idea yet that modern knowledge was superior to ancient; to the contrary, they tended to look to the past for authoritative knowledge.[13] Among alchemists this was particularly true because of a firm belief that ancient sages (*die alten Weisen*) had successfully created the philosophers' stone and that their wisdom had been preserved in the alchemical corpus. Thus, as repositories of an ancient wisdom that had wended its way over the course of a millennium from Hellenistic Egypt to Europe via medieval Spain and Baghdad, medieval Latin texts were an important resource for early modern alchemists.[14]

More than just conveying ancient knowledge, however, medieval authors also made their own contributions to the alchemical tradition. Natural philosophers writing in Arabic, such as Jābir ibn Hayyān (known in Europe as Geber, fl. 9th–10th c.) and Muhammad ibn Zakariyyā al-Rāzī (or Rhazes, ca. 854–925 or 935), digested and elaborated on Greek alchemical theory. In the process, they developed some of alchemy's central tenets, such as the idea that all metals are made of sulfur and mercury.[15] Following increased contact with the Muslim world, twelfth-century scholars in Europe began to translate Arabic natural philosophical and alchemical texts into Latin. These translations introduced Europeans to alchemy for the first time, and, in response, Latin scholars began to absorb, rework, and develop the vibrant medieval Arabic alchemical tradition.[16] Disagreement about the possibility of metallic transmutation and the limits of human technology took center stage in what William Newman has described as the "alchemical debate of the late Middle Ages."[17] As supporters of the alchemical arts such as the philosophers Albertus Magnus, Roger Bacon, and pseudo-Geber defended their art in the thirteenth and fourteenth centuries, they produced a number of influential texts dealing with alchemy's theoretical and practical elements. Alchemy's proponents ultimately failed fully to establish the art's legitimacy in the face of growing ecclesiastical and philosophical criticism during this period; nevertheless, these debates established a Latin textual tradition, a European alchemical corpus that continued to be influential into the seventeenth century.[18]

Alongside this Latin scholastic debate about alchemy's legitimacy, vernacular late medieval alchemical literature flourished as well. As Michela Pereira has noted, the very fact that the alchemical tradition "implied linguistic transfers from the beginning"—that is, that so many alchemical texts had already been translated from Greek into Arabic and then into Latin—made alchemists especially "likely candidates to use the vernacular in writing"; the notion of presenting alchemical ideas in a variety of languages, in other words, was deeply familiar. Moreover, alchemy had always operated in what Pereira calls a "double-regimen," oral and written, "that characterized alchemy in its dissoluble mingling of practice and theory." This double regimen ensured that the Latin and vernacular operated side by side in laboratory operations; practitioners who simultaneously read texts in Latin and spoke to each other in the vernacular might use both languages in books as well.[19] For all of these reasons, in the late thirteenth and early fourteenth centuries alchemical texts began to appear in French, Italian, Catalan, Czech, and English and in a wide range of genres, including recipes, poems, images, and prose treatises. Some were translations of Latin or Arabic texts, as was the German version of John of Rupescissa's book on the quintessence (ca. 1440), but other texts were originally written in the vernacular.[20] The German language alchemical tradition seems to have begun around 1415–19 with the *Buch der heiligen Dreifaltigkeit* (Book of the Sacred Trinity), which likened the process of creating the philosophers' stone to the death and resurrection of Christ.[21] Another German alchemical treatise appeared in 1426 as *Alchymey teuczsch,* which preserves the alchemical activities of four Bavarians, Niklas Jankowitz, Michael von Prapach, Michael Wülfling, and one Friedrich. Most famously, the illustrated alchemical poem *Sol und Luna* appeared in the late fifteenth century.[22] These texts emphasized an experimental alchemical tradition outside the universities and paved the way for the explosion of vernacular texts in the sixteenth century.

Whether Latin or vernacular, this medieval alchemical corpus was becoming increasingly available to sixteenth-century Europeans. Texts certainly continued to circulate in manuscript even a century after the invention of the printing press; indeed, until the middle of the sixteenth century, it is likely that most alchemical texts remained in manuscript form (and this is especially true if one takes into account recipe books).[23] When Sömmering recommended Sternhals's *Ritter Krieg* to Julius, for example, he could only have had a manuscript in mind, since this fifteenth-century text would not be available in print for another two decades. Nevertheless, from the middle of the sixteenth century onward printers seem to have discovered a healthy market for alchemical texts, particularly those that dealt with medicine and the transmutation of metals, since they issued sixteenth-century editions of older treatises attributed (both pseudonymously and authentically) to authorities such as Raymond Lull and Bernhard of Treviso.[24]

Sixteenth-century readers had access to ancient treatises on alchemy as well as works by medieval authors. Alchemical works erroneously attributed to Aristotle, for example, appeared in twenty-four print editions before 1536.[25] The most influential of the ancient alchemical sages to appear in print was Hermes Trismegistus, imagined to be a contemporary of Moses. As is now well known from Frances Yates's pioneering work, although some Hermetic writings (most prominently the Asclepius dialogue) were familiar to medieval scholars, Marsilio Ficino's translation of part of the Hermetic corpus into Latin in 1463 and subsequent publication of it in 1471 ushered in a new wave of enthusiasm for the writings of this ancient magus.[26] The Hermetic texts, which were actually composed in the first centuries CE by a variety of authors and were deeply influenced by Neoplatonism, inspired a revival of the mystical elements of alchemy in the early modern period. Those who pursued this side of alchemy emphasized its religious aspects, its role as a tool for understanding God rather than as one "simply" for making metals.[27]

Readers who wished to have a broader overview of the alchemical tradition could also buy compendia, or collected excerpts of authoritative texts. These collections drew on the much older manuscript florilegia tradition, in which a compiler gathered together quotations or excerpts from a variety of authorities, as well as on the more carefully organized Renaissance humanist commonplace books.[28] The first printed alchemical compendium, for example, the Nuremberg printer Johannes Petreius's *De alchimia* of 1541, included sections of pseudo-Geber's *Summa perfectionis* (ca. 1300), as well as Roger Bacon's *Speculum alchemiae* and the Hermetic *Tabula smaragdina.*[29] The compiler of the most well known of these printed collections, the *Rosarium philosophorum* of 1550, arranged quotes (at times quite lengthy) from, among others, Hermes Trismegistus, Arnald of Villanova, Aristotle, Democritus and Maria Prophetissa. The *Rosarium* literally placed these chronologically disparate authorities in conversation with one another, arranging excerpts of their texts into a dialogue on the philosophers' stone and accompanying it with a series of woodcuts to illustrate the stages of the alchemical work. Compendia such as the *Rosarium* distilled the vast alchemical tradition into a more manageable form, both in Latin and the vernacular German, thus making it more widely available to a literate public with a growing interest in alchemy.[30]

Sömmering's reading list demonstrates how print culture both preserved and invigorated this ancient and medieval alchemical tradition. As the sixteenth century neared its end, however, consumers of alchemical literature increasingly supplemented these older texts with more modern authorities. Thus, for example, when in 1578 the Bohemian nobleman and alchemist Bavor Rodovský mladší z Hustiřan (1526–ca. 1592) chose to translate into Czech what he deemed to be the most important alchemical literature, like

Sömmering he included Bernhard of Treviso, as well as the two key medieval alchemical compendia, *Rosarium philosophorum* and the *Turba philosophorum*.[31] Unlike Sömmering, however, Rodovský also included the work of a newer figure in the alchemical canon, Philippus Aureolus Theophrastus Bombastus von Hohenheim, known primarily as Paracelsus (1493–1541).[32] As Rodovský wrote in 1573, "Paracelsus is the most well-known philosopher and was certainly sent by Lord God to us on this earth." Accordingly, Rodovský felt it was his patriotic duty "to my beloved Czech homeland" to make Paracelsus's writings available in the kingdom of Bohemia.[33]

Rodovský's desire to translate Paracelsus into Czech is an indication of the Swiss reformer's importance as an alchemical authority in the decades after his death in 1541. Rejecting traditional humoral medicine and its accompanying therapeutic methods, Paracelsus advocated instead the use of drugs designed to attack the specific disease, rather than treat the balance of humors within the body as a whole. He also complemented the ancient doctrine of the four elements (earth, water, fire, and air) with the *tria principia* or *tria prima*, composed of the two medieval principles of matter (sulfur and mercury), as well as a third (salt). By simultaneously revolutionizing matter theory and resituating medicine on a new foundation of chemistry, Paracelsus and his followers gave chemical processes a new prominence in the study of nature generally.[34]

As Paracelsus's writings appeared in print and his followers began to digest and synthesize his work, his expansive ideas became all the more accessible. In the 1560s, Paracelsus's followers Adam von Bodenstein (1528–77), Michael Toxites (1515–81), and Gerhard Dorn (ca. 1530–after 1584) began to collect and publish his works, gradually establishing a corpus of Paracelsian texts. Dorn, in particular, played a key role in identifying, publishing, and interpreting (sometimes also falsifying) Paracelsian manuscripts.[35] The Danish physician Petrus Severinus (ca. 1542–1602) also contributed to the efforts to make Paracelsus's vast writings comprehensible. In his 1571 *Idea medicinae philosophicae* (Idea of Philosophical Medicine), Severinus synthesized Paracelsus's work and distilled from it a coherent medical system comparable (in his view) to Galen's and Aristotle's.[36] The mid-1570s saw the publication of three "onomastica," or thematic lexica, of Paracelsus's unique terminology by Michael Toxites, Leonhard Thurneisser, and Adam von Bodenstein.[37] Finally, in 1589–90, Johannes Huser (before 1545–ca. 1600) issued Paracelsus's collected works.[38] This first generation of Paracelsians and their efforts to explicate and publicize the reformer's ideas were essential to his success and prominence as the seventeenth century began.

Paracelsus's place in the pantheon of alchemical authorities was always ambiguous for those who prized alchemy primarily for its promise to transmute metals. Although Paracelsus used the term "alchemy" repeatedly in his writings, he used it primarily in the context of chemical medicine, which came to be called *chymiatria* or *iatrochemistry*. He sought to reorient the art away from the production of gold and toward medicine, disparaging alchemists who sought *only* to transmute metals. In his *Paragranum*, Paracelsus stated this explicitly, writing that alchemy's purpose is "not, as they say, to make gold, to make silver; here [the purpose] is to make arcana [or medicines] and apply them to illnesses."[39] But Paracelsus was never entirely consistent in his voluminous writings, and readers could easily find places where he seemed to support the possibility of producing alchemical gold. In his treatise on the generation of metals, *De mineralibus*, for example, he discussed the alchemical aim of perfecting or even surpassing nature (including metals) via human skill. Here he repeated many commonplaces of transmutational alchemy, even commenting that "transmutation can happen in alchemy."[40] As Joachim Telle has noted, "in *De mineralibus*, one finds no sign of that Paracelsus, present elsewhere, who sought to defeat the fundamental beliefs of alchemists active in the transmutation of metals."[41]

Alchemists interested in producing gold thus could find support for their ideas in the Paracelsian corpus, even if Paracelsus disparaged transmutational alchemy elsewhere in his writings. "Goldmakers" could also find legitimation in the reformer's growing reputation as the "German Hermes Trismegistus," holder of the secret of transmuting metals.[42] A number of spurious stories began to circulate soon after Paracelsus's death, spreading the belief that he had actually possessed the philosophers' stone, and—following the venerable alchemical tradition of pseudonymous texts—books on transmutation appeared in his name or were attributed to him. Thus, for example, an anonymous treatise from around 1550–70 that was among the most popular alchemical tracts in German, "Vom Stein der Weisen" (On the philosophers' stone), was attributed to Paracelsus already in 1575.[43] Even Paracelsians were guilty of propagating such pseudonymous texts. Adam von Bodenstein and Michael Toxites, two of Paracelsus's early editors, issued a number of spurious alchemical tracts in his name, including *Liber vexationum* (1567), *Manuale de lapide philosophorum* (1572), *Thesaurus thesaurorum* (1574), and *Tinctura physicorum* (1572).[44] What historians have since, with hindsight, identified as Paracelsus's clear rejection of transmutational alchemy was far from clear to his followers in the first century after his death. From wild rumors and spurious texts to the words of the reformer himself, early modern alchemists had plenty of reasons to see Paracelsus as an authority on both medicinal *and* transmutational alchemy.

The literature of Paracelsus and his followers had a substantial impact on the alchemical tradition. To practical, mystical, and natural philosophical alchemical texts of antiquity and the Middle Ages Paracelsianism added a new emphasis to the alchemical arts: chemical medicine or iatrochemistry. The sixteenth century also produced another type of textual resource for students of the alchemical arts: skills booklets or *Kunstbücher*. As William Eamon has shown, this new genre, created largely by printers, made practical, "how-to" knowledge accessible to craftspeople and laypeople alike in the sixteenth century.[45] These *Kunstbücher* divulged the secrets of illuminating manuscripts, dying and cleaning fabrics, and working with metals. An indication of alchemy's widening appeal, one early *Kunstbuch* even offered its readers alchemical arcana gleaned from a "famous alchemist from Mainz," Petrus Kertzenmacher. Kertzenmacher's booklet, which was issued (and edited differently) by two different printers, instructed his readers primarily in the technical aspects of alchemy, disclosing the techniques, materials, and equipment needed for the most basic alchemical operations.[46] Among "the things that belong to the art [of alchemy]" included in the book were recipes for the preparation of substances essential to the alchemist's craft, including cinnabar (mercuric sulfide), *Spangrün* (a salt from copper), *Weinstein* (tartar), and *aqua fortis* (or nitric acid, used in dissolving gold).[47] Kertzenmacher taught his readers how to use these substances as well, outlining, for example, how to make silver out of sulfur and quicksilver.[48] Perhaps most important, the two printers who issued versions of Kertzenmacher's book did so in a format that was easy to use. The book contained numerous illustrations of stills and furnaces, as well as an index to facilitate finding specific recipes. (See fig. 1.) The book even included a translation into German of frequently used Latin words with which a less well-educated audience may not have been familiar, such as *sol* (sun, or gold) and *luna* (moon, or silver). Armed with this kind of informative book and the requisite materials, someone literate in the vernacular German, if not Latin, could easily begin his or her training in what Kertzenmacher promised was "the highest [art] of them all."[49]

Sixteenth-century students of alchemy thus had an extraordinarily rich and diverse textual tradition to which to turn. Medieval authors provided both sophisticated discussions of alchemical theory and material on practice, while an ancient thread, revived in the Renaissance, stressed the more

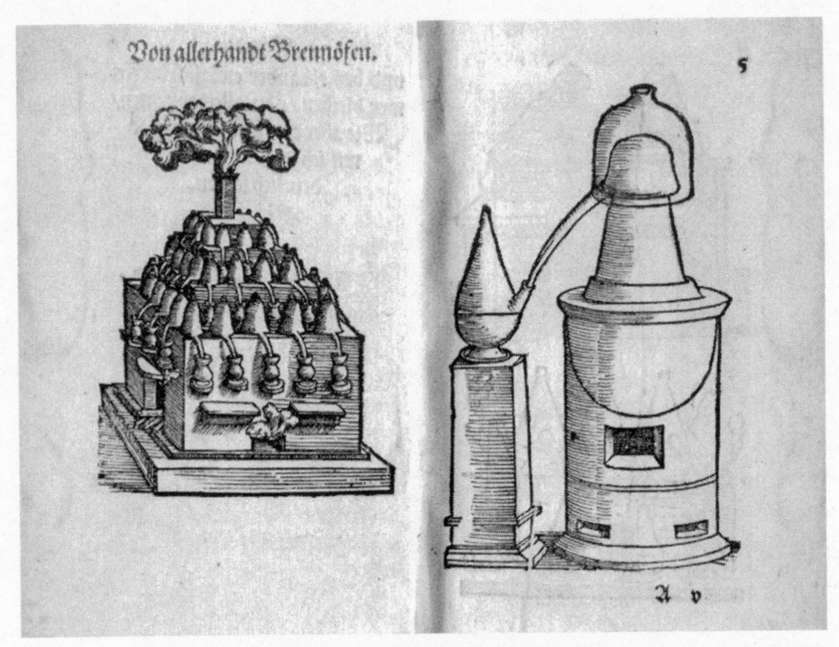

Figure 1. Distilling furnaces, as depicted in [Peter Kertzenmacher], *Alchimia das ist alle Farben, Wasser, Olea, Salia, und Alumina . . . zubereyten* (Frankfurt am Main: Bey C. Engenoffs Erben, 1570). (From the Roy G. Neville Historical Chemical Library, a collection in the Othmer Library, Chemical Heritage Foundation. Photo by Douglas A. Lockard.)

mystical elements of the alchemical art. As the Paracelsian corpus came together, it began to offer a new framework for the practice of alchemy as chemical medicine. Finally, *Kunstbüchlein* such as Kertzenmacher's treatise made alchemical knowledge accessible to a lay audience, one not fluent in the Latin of scholars but nonetheless interested in understanding and manipulating nature.

In fact, Renaissance readers could find so many texts about alchemy that they were confronted with what Ann Blair has called "information overload."[50] Like many alchemists, Philipp Sömmering was keenly aware of this problem. He advised his patron to wade carefully through the bewildering variety of alchemical texts, comparing each book with the others and keeping track of the points of agreement: "He who has them [i.e., good books], should then take care to read them in a certain order. And at all times make a note of where the philosophers disclose one or more secrets. Because where most of them agree [*concordiren*], there they speak the truth. Because no one says or sets out all secrets, but rather one explains this, another explains something else. This is why they say *the book explicates the book* [*liber librum explicat*]. Because then, Your Grace, when you have gotten all of the most outstanding books, in which the foundation and the truth of this art will be taught, then I wish to offer you most humbly my modest instructions as to how you should read one after the other."[51] This aphorism, *liber librum explicat* (or *aperit*), was a commonplace in alchemical literature, emphasizing the importance not just of books, but also of reading them properly.[52] Not all alchemists took the time to read so carefully and thoroughly, however. For every well-read alchemist like Sömmering, there were countless more who owned only one or two books, from which they hoped to distill the entire truth of alchemy. The presence of manuscript compendia only compounded the tendency to rely on one book, albeit with several different texts inside. With such a diversity of books to choose from, alchemical practitioners thus came to a variety of truths, all of which had roots in the vast textual tradition available to sixteenth-century readers.

Adepts and Alchemical Illumination

A sixteenth-century alchemist looking to this tradition, however, would quickly have seen that books were not the sole source of alchemical wisdom. Divine illumination, as well as the oral transmission of secrets through initiation and scholarly training at the side of a master, had an important role in alchemical lore and literature. Although it is unclear whether these methods were merely a rhetorical trope or the real practice of ancient and medieval alchemists, such anecdotes suggested to early modern students of alchemy that they, too, might learn the secrets of the art not just from books, but also from God or other adepts. Both paths to wisdom emphasized the ancient idea that alchemy was privileged knowledge, a sacred art to be divulged only to the few.[53]

The notion of alchemy as *donum dei*—a gift from God to a select group of people—was of ancient provenance.[54] Maria Prophetissa (before the early 3rd century CE) believed that God gave alchemy specifically to the Jews as his chosen people.[55] In Byzantine and medieval Arabic alchemy, the idea of alchemy as *donum dei* shifted. God no longer bestowed his gift on a particular religious group or people but now on a group of adepts chosen to receive such wisdom. Although it is unclear how such divine illumination really happened and to whom, this formulation of the *donum dei* persisted in medieval and early modern European texts. Thus, the Silesian alchemist Christoff von Hirschenberg (fl. 1574–92) invoked this tradition in a fairly typical fashion in 1583 when he asked his patron for "a comfortable, quiet and safe place so that with . . . protection I might finish this gift given to me from Almighty God."[56]

For those who did not receive alchemical wisdom directly from God the alchemical tradition offered the alternative model of studying with a master. As Sternhals explained in the *Ritter Krieg*, many learned men had tried to understand alchemy by reading, only to fail, sometimes at great cost, because the art "is not of the world, but can be understood only from the true sons [of alchemy], or from those to whom they have revealed and taught the exalted secret in person out of particular favor and inclination."[57] According to this notion, just as God passed the sacred art of alchemy to the chosen adept, so too might the adept hand down the secrets to a worthy novice. As the English alchemist Thomas Norton (ca. 1433–1513/14) put it in verse in his 1477 *Ordinall of Alchimy*, the adept had a duty, in fact, to pass on God's gift to a single successor: "Of the person which shall this *Scyence* leere, / And in likewise make him straightlie swere: / Soe that noe man shall leave this *Arte* behinde, / But he an able and approved Man can finde; / When Age shall greeve him to ride or goe, / One he may teach, but then never no moe."[58] According to legend, this transfer of divine wisdom was to occur face-to-face, never in writing, and traditionally was accompanied by vows of secrecy.

When Norton narrated the story of his own education in the *Ordinall*, he drew on all of these elements to create a classic tale of alchemical initiation (fig. 2). Knowing that alchemy's true secrets could only be had from an adept, Norton decided to seek out a master and wrote to the famed alchemist George Ripley (ca. 1415–ca. 1490). Ripley eventually responded (according to Norton), impressed with Norton's virtue and perseverance. "My very trusty, my deere beloved *Brother*," he allegedly wrote, "I must you answer, it may be none other; / The tyme is come you shall receive this Grace, / To your greate comfort and to your solace . . . / Wherefore it is needed that within short space, / Wee speake together, and see face to face: / If I shulde write, I shulde my fealty breake, / Therefore Mouth to Mouth I must needs speake."[59] Next, Norton recalled,

> This *Letter* receiving, I hasted full score,
> To ride to my *Master* an hundred miles and more;
> And there Forty dayes continually,
> I learned all the seacrets of *Alkimy* . . .
> There fownd I disclosed the *Bonds of Nature:*

Figure 2. An alchemist receiving the *donum dei,* as depicted in Thomas Norton, *The Ordinall of Alchimy* in *Theatrum chemicum britannicum,* edited by Elias Ashmole (London: Printed by J. Grismond for Nath. Brooke, 1652). (Beinecke Rare Book and Manuscript Library, Yale University.)

The cause of Wonders were to me soe faire,
And so reasonable, that I could not dispaier.
If your *master* and ye resemble all aboute
My good *Master* and me, than have ye no doubte.[60]

From its insistence on face-to-face instruction in the "Science of Holy Alkimy" to the oath of secrecy, Norton's story articulated the traditional path to alchemical enlightenment and served as an ideal for his sixteenth-century readers.

Both the idea of alchemical knowledge as *donum dei* and the idea that alchemical learning could be acquired directly from the mouth of an adept rested on the assumption that the alchemist could not be self-taught, but must be *chosen* to receive the alchemical arts. As Norton put it, "Also no man coulde yet this *Science* reach, / But if *God* send a *Master* him to teach."[61] Nor, in theory, was alchemy to be passed on through the family, as were the crafts. Norton insisted that the adept "shall not be so wilde / To teach this secreat to his own childe; / For nighnes of Blood ne Consanguinity / May not be accepted to this dignity: / Soe blood as blood, may have hereof noe part, / But only vertue winneth this holy *Arte.*"[62] Legendary stories like Norton's reinforced the idea that alchemy was knowledge reserved only for the very few, and, moreover, that those few must be chosen; in other words, one could not simply decide to become an alchemist.

The Alchemical Peregrinatio

Norton's story, however, was composed of the stuff of legend; few alchemists in the early modern period (if ever) actually had this kind of alchemical training. In a sense, Sömmering tried to position himself as the master who might reveal alchemy's divine secrets to Duke Julius. When he recounted the details of his own alchemical education during his trial in 1574, moreover, he too included face-to-face interactions with more knowledgeable practitioners, but his encounters were of a very different sort from Norton's a century earlier. After he bought his first two books on alchemy in the late 1550s and resolved to study the art, Sömmering recalled that he went to Erfurt. There he paid a wood-carver five thaler to teach him two of alchemy's most basic operations: distillation and sublimation. In the same city he visited an apothecary's shop, where he diligently wrote down "all species" of plants and purchased eleven thalers' worth of herbs, then discussed alchemy with a "Philosophus" named Georg von Eidenbach. Sömmering evidently continued his alchemical education in the local tavern, for, according to his testimony, "from an alchemist named Martin Gurlach, he learned an entirely secret art while drinking, namely, the *regimen ignis*," or the art of regulating the intensity of the fire in alchemical stills and furnaces.[63]

Sömmering's tale clearly does not have the mythical quality of Norton's instruction at the side of a great adept. In part, this difference is certainly due to the fact that Norton's story appears in a treatise written for public consumption (in rhyming couplets, no less), while Sömmering's was extracted from him at the beginning of a trial that would lead him to the scaffold. But Sömmering's description was certainly more typical of the way most practitioners learned to be alchemists in the sixteenth century. He did not have the privilege of an adept's tutelage. With a confidence that certainly annoyed alchemists who shared Norton's more privileged view of alchemy, Sömmering did not wait to be chosen, but simply chose himself; he evidently felt that he could understand alchemy on his own by cobbling together an education from a variety of people, just as readers did with texts. The fact that philosophers, other alchemists, apothecaries, and even unlikely sources like the wood-carver all seemed to have had learning and skills useful to Sömmering is itself an indication of how widespread the elements of alchemical knowledge were by the mid-sixteenth century.

This type of piecemeal collection of useful skills and techniques was common among alchemists of all backgrounds. Paracelsus himself, in fact, was the best-known advocate of gathering wisdom from a variety of sources. Playing on the guild practice of the journeyman's years of travel and training, Paracelsus claimed that during his own *Wanderjahren* he had sought out the "the true art of medicine" not only from learned doctors, but also "from barbers, bathers, learned doctors, wives, those who make a habit of black magic, from alchemists, at cloisters, from nobles and commoners, from the clever and the simple."[64] This model of education—from folk as well as learned sources—was profoundly influential for later practitioners of both medicine and alchemy. The alchemist Christoff von Hirschenberg probably had Paracelsus's travels in mind when he appealed to the Bohemian magnate Vilém Rožmberk for patronage in 1583. In his carefully constructed narrative about his training in the alchemical arts, Hirschenberg told Rožmberk, "As soon as I finished my modest studies, I began to wander through various lands, during which time in truth I kept company not only with many learned men I came across, but I also learned something from a man from another village."[65]

Princes themselves were unlikely to travel the Holy Roman Empire in search of bits of alchemical wisdom, but they could collect experienced alchemists who had. This wish to learn—to take "alchemy lessons," in a sense—was an important part of the princely interest in alchemists in the sixteenth and seventeenth centuries. Patrons could discuss alchemy with the practitioners they hired, and had the opportunity to drop in and observe laboratory work in progress. Some sovereigns arranged tutelage more formally in contracts, stipulating that an alchemist teach as a part of his or her employment. Duke Johann Friedrich of Württemberg (1582–1628) and his mother Sibylla (1564–1614), for example, signed such a contract in 1610 with three alchemists named Andreas Reich, Johann Andreas Hess, and Sebastian Hesch. The trio promised to teach the royal family "both in writing and orally" "the fundamentals and principles of the Art of Alchemy (by which the tincture or philosophers' stone, or the *principium movens*, naturally transmutes quicksilver and other imperfections to a perfect metal, namely to gold and silver)."[66] Promising a complete education, the alchemists assured the duke and his mother that they would learn both "the theoretical and practical knowledge [*scientia*] with all demonstrations and manipulations, but in particular the preparation of the essence of gold and silver, and also the entire fundamentals of the tincture, in addition to the essences of spirits, sulfur, arsenic, and quicksilver" so that "Your Graces will know how to prepare it."[67]

In some ways, contracts like these echoed the mythical adept-novice relationship of alchemical tradition, but they also transformed it. Norton believed that alchemy was to be reserved for the select few, the virtuous and the humble.[68] "Of a *Million*, hardly *three* / Were ere Ordaind for *Alchimy*," he wrote.[69] Reich and his colleagues, on the other hand, were more generous and did not hesitate to share their expertise with the ducal family of Württemberg. These contracts also reflected a shift in the accessibility of alchemy. It was still considered a honorable and prized art, but by the sixteenth century, princes and practitioners seeking alchemical arcana no longer had to rely on the good fortune of finding that rarity, the alchemical adept, nor did they have to wait for God to bestow the gift upon them. In part because of the diffusion of alchemical knowledge through books, aspiring alchemists in the sixteenth century could find plenty of people willing to share their knowledge—usually for a price. In this new marketplace of secrets, alchemy was available to princes, philosophers, physicians, and pastors—in short, to anyone with the money and enthusiasm to seek it out.

Alchemy as a Craft

Alchemy was particularly accessible to practitioners who already had skills and knowledge from related and overlapping fields. Crafts such as goldsmithing and medicine, for example, were intimately related to alchemy in the skills they employed and the knowledge they required. Likewise, a general background in philosophy provided a framework for alchemical theory and aided the comprehension of often deliberately confusing alchemical texts and erudite references. Not surprisingly, then, many practitioners came to alchemy with training or previous experience in allied fields in the artisanal or scholarly study of nature.

Distinguishing alchemy from metallurgy (broadly speaking) and medicine in the early modern period is quite difficult. As the Italian mining expert Vanoccio Biringuccio pointed out in 1540, the fields of alchemy, distilling, goldsmithing, and metalworking were kindred arts in that they were all "works of the fire."[70] From the practitioner's perspective, this meant that these fields of expertise offered a vast reservoir of knowledge that could easily be parlayed into more explicitly alchemical tasks. The career of Leonhard Thurneisser best exemplifies the way practitioners could do this.[71] Born in 1531, Thurneisser followed in his father's footsteps and trained to become a goldsmith. By 1559, Thurneisser had begun to work in mining, overseeing a mine in Terrenz in the Tyrol. His success there with a smelting and sulfur works drew the attention both of learned men and of Archduke Ferdinand II of Austria. When the archduke hired Thurneisser, he sent him on numerous travels and research trips to study metallurgy, during which Thurneisser also learned a great deal about Paracelsian medicine. Returning to the Tyrol to find the mines in serious disrepair, Thurneisser left behind his career in mining for one in chemical medicine. Thurneisser's publication of three books on various aspects of *chymiatria* from 1569 to 1571, *Archidoxia, Quinta Essentia,* and *Pison,* gained him notoriety. Shortly thereafter, Elector Johann Georg of Brandenburg called him to Berlin, where Thurneisser then took up a post as the electoral physician. Just before leaving this position in the face of growing criticism (and perhaps in an effort to keep his job), Thurneisser published his only treatise on transmutational alchemy, *De transmutatione veneris in solem* (On the transmutation of Venus [copper] into Sun [gold]).[72]

Thurneisser was able to transform himself from goldsmith to mine operator, to Paracelsian physician, and finally to alchemist because these various operations all shared similar vocabulary, materials, tools, and skills.

Practical experience like that Thurneisser gained in his father's goldsmith's shop or in the mines could be put to use in the context of alchemy. As the early seventeenth-century learned alchemist Michael Maier put it, "the work of the goldsmith and of other metalworkers must not be unknown to [the would-be alchemist]."[73] Skill in distilling was useful in alchemical work as well.[74] The medical practitioner Georg Hört, for example, turned his medical distilling skills to transmutational alchemy in 1595.[75] Hört was not an alchemist by trade, but rather claimed that he was "a Theophrastan *medicus,* [who] cures all kinds of illnesses with a Theophrastan art and medicine, distills and makes waters, oils, and the like."[76] When Hört's acquaintance David Pirkheimer asked for assistance in making gold for Duke Friedrich I of Württemberg, Hört claimed that the invitation came because he "was capable with the fire and distillation."[77] The numerous mining regions of central Europe provided particularly fertile ground for training future alchemists.[78] Assayers, in particular, were not only trained to recognize various ores, extract precious metals from them, and analyze samples to determine their contents; they were also highly skilled in the use of the fire to work with metals and in building and operating crucibles, furnaces and tools.[79] All of these skills were indispensable to the alchemist as well, so much so that Maier remarked that the art of the assayer was "not only extremely useful for the future *chymicus,* but absolutely necessary."[80]

Would-be alchemists who did not have firsthand experience in mines or apothecary shops could turn instead to a burgeoning literature on distilling and metallurgy. In the late Middle Ages, Franciscan friars such as pseudo-Lull, Roger Bacon, and John of Rupescissa had developed a fundamental corpus of works on distillation as a means to create a potent medicinal elixir, sometimes known as the quintessence, or *aqua vitae,* out of any number of mineral and organic substances. In the early modern period, these authors' works appeared in print in Latin and in translation. The *Coelum philosophorum; seu, Secreta naturae* (Heaven of the philosophers, or, secrets of nature, 1543), for example, collected the works of pseudo-Lull, pseudo-Arnald, Rupescissa, and others on how to distill the quintessence, or *aqua vitae,* "not only from wine, but also from various metals, fruits, meats, eggs, roots, herbs, and many other things."[81] The 1576 *Herliche medicische [sic] Tractat, vor nie in Truck kommen* (Wonderful medicinal treatise, never before in print) offered similar knowledge in German, but added several sixteenth-century German texts to the medieval classics.[82] Indeed, the sixteenth century had seen a proliferation of new vernacular works on distillation, including most famously Hieronymus Brunschwig's *Liber de arte distillandi* (Book on the art of distilling, German edition, Strasbourg, 1527) and the *Destillier Buch* of Walther Hermann Ryff (Distillation book, Frankfurt am Main, 1545). These practical distillation texts not only included accessible, "how-to" information on distillation; they also forged a crucial connection between alchemy and medicine by emphasizing alchemy's function as the separation of the pure from the impure or even poisonous.[83]

Mining and practical metallurgical literature was also a rich source in the sixteenth century, for, as Pamela Long has noted, in central Europe this was "the great age of mine and metallurgical literature both in terms of quantity and originality."[84] The first printed book on mining, physician and mine investor Ulrich Rülein von Calw's *Bergbüchlein*, appeared anonymously about 1500[85] and was followed in the 1520s by another practical booklet on assaying, the *Probierbüchlein*.[86] These two early booklets were part of the same *Kunstbuch* tradition as Kertzenmacher's *Alchimia* in that they aimed to instruct both practicing miners and nonexperts in essential mining and assaying techniques. Although the *Bergbüchlein* deals more with the generation and growth of metals and the *Probierbüchlein* contains recipes for assaying metals, both conveyed practical information alongside more theoretical knowledge.[87]

These early pamphlets were followed by two more substantial scholarly central European treatises on mining and assaying: Georgius Agricola's *De re metallica* (On metals, published posthumously in 1556) and Lazar Ercker's *Beschreibung allerfürnemsten mineralischen Ertzt unnd Berckwercksarten* (Treatise on ores and assaying, 1574).[88] Agricola aimed at a systematic and sweeping review of the knowledge of metals, as well as the techniques and tools used to mine and work them, whereas Ercker's treatise focused more narrowly on assaying. Both authors expressed their skepticism about transmutational alchemy in these printed works, but neither could ignore the contributions of the alchemists altogether. As Agricola and Ercker grudgingly admitted, in the absence of a learned discipline of metallurgy before the sixteenth century, alchemical literature, together with ancient Greek philosophy, was the only repository of knowledge about metals. Thus, for example, Agricola acknowledged that "the alchemists have shown us a way of separating silver from gold by which neither of them is lost," and Ercker granted that "assaying is a very excellent, ancient, and useful art which, like all the other arts that work with fire, originated very long ago as an outgrowth of alchemy."[89]

Together, all of these books imparted a great deal of practical metallurgical knowledge to their readers. Although the authors had different backgrounds and levels of education, Pamela Long has argued that this literature all "shared a common context that included the early modern capitalist expansion of mining."[90] Because these authors aimed in part to educate potential investors in the methods and processes of German mines and metallurgy, they valued openness and clarity in their writing. Long has described how they shared an appreciation of "clear technical language and understandable discussions of technical processes, careful measurement, honest and precise assaying, and practical skill," all of which made their writings particularly useful and accessible to anyone interested in alchemy as well.[91] Especially useful to alchemists were these authors' discussions, accompanied by helpful illustrations, of equipment and processes. Ercker, for instance, explained "how to make and equip assay furnaces" and various clay vessels, as well as how to regulate the fire (that most prized art which Sömmering picked up from an alchemist in a tavern), all of which would have been extremely useful to an alchemist who did not yet have or know how to make and use the proper equipment to practice his or her art.[92] Ercker also explained how to make various "waters" or acids, like *aqua fortis,* which alchemists, goldsmiths, and assayers all used to separate and dissolve metals.[93] A particularly striking feature of these books was the illustrations that accompanied the explicative text (figs. 3, 9, 10, 11). Not only did authors like Ercker explain how to make clay vessels and crucibles, but they also provided pictures of the finished product that would have helped any reader trying to follow instructions.

The mining and metallurgical literature of the sixteenth century was thus a rich source of information for alchemical practice, or the actual work of separating, amalgamating, and refining metals. These books were less helpful, however, as a resource for alchemical theory, or the larger philosophical context in which many (although not all) practitioners operated; for this, some practitioners benefited from a broad training in natural philosophy and other fields of knowledge. Comprehensive knowledge of the scholarly writing on nature not only provided a deeper understanding of the underlying principles of alchemical processes, but also helped novices understand the often confusing alchemical literature. The more scholarly alchemists, not surprisingly, maintained that anyone who lacked such training in natural philosophy could never hope to become a real alchemist. As the alchemist and physician Michael Maier noted, "sensible operation or manual experimentation . . . is blind without previous theory"; therefore, the true alchemist must have not only a familiarity with mining and metallurgy, but also a rigorous education in the liberal arts.[94]

Figure 3. Distillation and assaying apparatus, as depicted on the title page of Lazarus Ercker, *Beschreibung allerfürnemisten mineralischen Ertzt unnd Berckwercksarten* (Frankfurt am Main: Georg Schwartz, 1580). The man in the center holds a cucurbit, alembic, and receiver, while a Lazy Heinz furnace rests to his right surrounded by other distillation equipment. The furnaces at the upper right and left would have been used for assaying. (From the Roy G. Neville Historical Chemical Library, a collection in the Othmer Library, Chemical Heritage Foundation. Photo by Douglas A. Lockard.)

Poetry, grammar, rhetoric, and logic, he claimed, were necessary even to be able to read alchemical texts properly. Because they often utilized a highly symbolic, if not deliberately obfuscating, language, alchemical texts could be very complicated. Thus, Maier argued, these literary skills were the most important of all because "they form the basis for all of the rest."[95]

Maier required an understanding of geometry, arithmetic, astronomy, physics, and medicine in part for the same reason. Because they cited other authorities like Aristotle so often, the writings of learned medieval alchemists would have been difficult to understand without a grasp of natural philosophy. As Maier argued, "from these arts and sciences [alchemical] authors chose all of their terminology and allegorical descriptions, all of which will remain incomprehensible and obscure to someone seeking the hidden truth in them, if he is not well-versed in them. How often do they speak of heavenly signs, of the motion of the sun and moon, of eclipses and the like? How often of the *materia prima* and reductions to it, of generation and corruption, of the seeds of plants and animals and innumerable things of this sort?"[96] Natural philosophy, in other words, would make alchemical texts comprehensible, but it had additional benefits as well. Because alchemy drew heavily on all other branches of knowledge, Maier suggested, one could simply not understand it in isolation. According to Maier, "indeed, in all of physics or nature, in all of medicine there are no precepts or principles, no causes or effects which the science of Chymia does not borrow and take for itself."[97]

Maier pointed out forcefully how critical a broad base of learning was for an alchemist; certainly many alchemists developed their interest in alchemy out of a comprehensive study of natural philosophy. It is important to recognize, however, that not all of Maier's contemporaries came to alchemy with this kind of training. Indeed, plenty of practitioners were quite successful without it, just as others claimed to be alchemists without the firsthand experience in assaying that Maier demanded. Indeed, mining provides an analogous case. In *De re metallica,* Agricola claimed that the "many arts and science of which a miner [i.e., mine investor] should not be ignorant" included philosophy, medicine, astronomy, surveying, arithmetic, architecture, drawing, *and* the law. Agricola's reasons for each of these were sound enough. Arithmetic, for example, was necessary so that the mine operator "may calculate the cost to be incurred in the machinery and the working of the mine."[98] But surely not everyone who managed central Europe's mines was accomplished in all of these areas. By the same token, knowledge of mining, metallurgy and natural philosophy were useful, but not essential, to students of alchemy, particularly given the many other resources for alchemical knowledge upon which they might draw.

This point can be extended to alchemy as a whole. There was no single trajectory or even a prescribed way to alchemical wisdom. Individuals could become alchemists through books alone, or in the tavern, the mines, and the pharmacy. Of course, some alchemists believed that there *was* a single path, that, for example, only those trained in natural philosophy were true alchemists. In the late sixteenth and early seventeenth centuries, those with access to print became increasingly vocal about these beliefs, issuing treatises in order to define just who could become an alchemist.[99] Yet clearly there was a wide variety of practitioners of alchemy, with diverse backgrounds and varied goals. More important, there were no institutions to regulate the practice of alchemy and the training of its practitioners, as there were in other trades and scholarly pursuits. The market for alchemy—in terms of the consumption of alchemical techniques and secrets—supported all of its varieties, from the pharmacist offering to distill an *aqua fortis* for an alchemist friend to the most learned natural philosopher's treatise. It was this tension—between the sheer variety of alchemy supported by the marketplace and the more restrictive definition of alchemy that individual practitioners tried to impose—that would create battles over definitions of alchemy at the end of the sixteenth century.

CHAPTER TWO

The Alchemist's Personae

Pieter Brueghel the Elder's *The Alchemist* (ca. 1558) presents the viewer with a chaotic scene (fig. 4). To the left a shoddily clad figure sits at the fire, dropping a coin into a crucible while various other materials smoke and bubble in the vessels and distilling apparatus around him. Above his head, pinned to the chimney, is a piece of paper with the word *misero* written at the top. To the right, a scholarly figure sits at a lectern, his outstretched hand gesturing at the scene before him, while with his other hand he points to a book. In large letters at the top of a page we see the words ALGE MIST, a pun meaning *alchemist,* of course, but also "all failed" (*al ghemist*) in Flemish or even "everything's dung" (*alles Mist,* perhaps, more colloquially, "everything's crap") in German. In front of the scholar on the floor crouches a man in a fool's cap, puffing the bellows at an overturned bunch of crucibles in a small brazier. Between the fool and the scholar rest a number of other tools and vessels. The rest of the scene is taken up with the destitute family of the figure at the fire on the left; inside, the wife shows the viewer her empty purse, while the couple's children jump into a cupboard conspicuously empty. Out the window we see the family's sad future, as they are received at a hospital in search of charity. Clearly the coin the father dropped into the crucible was their last, and it did not yield the wealth he hoped for.

As the various Latin, French, and Flemish inscriptions printed with *The Alchemist* made clear, Brueghel (1525–69) offered early modern viewers a morality tale of the futility of alchemy, warning those who were tempted that it could only destroy families and households. "See how this foolish man distills in his vials / The blood of his children, his treasures and his senses," read one French inscription. "See how, after searching uselessly for mercury, / He goes with his children to seek his bread."[1] The moral of the story seems clear. Yet as a representation of the alchemist, Brueghel's

Figure 4. Engraving by Philips Galle after Pieter Brueghel the Elder, *The Alchemist*, Antwerp, ca. 1558. (Philadelphia Museum of Art: The Muriel and Philip Berman Collection, 1985.)

image is far more indeterminate. Who exactly is the alchemist in this image? Is it the man at the fire, the father who has driven his wife and children into the poorhouse? What, then, of the fool, whose position at the fire suggests that he too is posing as an alchemist, or the scholarly figure, who is also surrounded by alchemical books and tools? Are they not also alchemists, perhaps of a different sort from the destitute father? Brueghel's sympathies in this image are equally unclear. Perhaps he wished his viewers to identify with the scholar, who, slightly removed from the rest by his anachronistic dress and positioning, seems to be passing judgment on the scene in front of him; perhaps he is meant to represent a more legitimate kind of alchemical practice than the foolishness going on around him. On the other hand, perhaps this same scholar bears responsibility for the family's fate. We could also imagine him directing these alchemical operations, reading out instructions to one or both practitioners before him, just as professors of medicine had long directed anatomies carried out by surgeons before them. Or perhaps the scholar is more deceptive, beguiling the alchemists before him with the worthless knowledge in his books and driving the man and his family to the poorhouse. However clear Brueghel's moral with respect to alchemy, he offers only an indeterminate representation of the alchemist: was he an honest—if greedy—*Hausvater*, a bumbling fool, or a scholar? Or perhaps all three?[2]

Given the variety of ways in which practitioners could access alchemical secrets in the sixteenth century, it was unavoidable that alchemists might be understood to be least as diverse as the three central figures in this image. Some, for instance, saw in alchemy a stimulating body of literature and a set of practices that could be used to explore fundamental problems in natural philosophy. Others saw alchemy as an extension of other artisanal practices, like metalwork, mining, or the production of drugs and medicines. Some alchemists focused primarily on the coveted philosophers' stone, to be sure, while still others found alchemy most useful for its rich metaphorical resonances. Alchemists' approach to their art differed for many reasons, including social, intellectual, and cultural positioning. As each individual put his or her own stamp on alchemical study and practice, the result was a proliferation of alchemies in the Holy Roman Empire that could appeal to many different audiences.

The activities of individual practitioners, however, were only partially responsible for shaping alchemy in the sixteenth century. For alchemy to be meaningful as a social practice, rather than just as a private activity, its practitioners needed to develop a recognizable identity as *alchemists* that could mediate between individuals and the society and culture in which they lived and practiced. Lorraine Daston and H. Otto Sibum recently have suggested the concept of the scientific persona as a fruitful way to think about this problem in the history of science generally. Drawing on a 1938 essay by the French anthropologist Marcel Mauss, they suggest, "Intermediate between the individual biography and the social institution, lies the persona: a cultural identity that simultaneously shapes the individual in body and mind and creates a collective with a shared and recognizable physiognomy." Daston and Sibum draw particular attention to the original Latin meaning of *persona* as mask. In contrast to the modern sense of mask as a "topos of insincerity" that disguises a more genuine self, Daston and Sibum point to the ancient meaning of the mask, "the dramaturgy of masks as makers, not destroyers of true identities." They insist, in other words, on the productive work of the persona, its function as an interface "between the society that must grant significance to a persona and the individuals who must embody it," and on "the hybrid character of the persona concept between individual and society. Symbols, values, and meanings—the stuff of culture—are essential components in this interaction."[3]

The emergence of a persona was a crucial component of alchemy's development in late medieval and early modern Europe. This persona provided an interface between a specialized theoretical and practical activity, on the one hand, and society, on the other; it embodied alchemy and gave it social and cultural meaning. But where did this persona come from? In part, alchemists assembled it themselves, taking elements from a number of other relevant personae in a kind of "cultural cut-and-paste."[4] They certainly had a number of possible choices. Alchemists could model themselves on scholars and cultivate an emotional detachment, as Gadi Algazi has described it, retreating into their studies (or laboratories) to escape the troubles of the world.[5] They might also have found appropriate models in the personae of artisans, merchants, or courtiers. Alchemists seemed to have an array of possibilities out of which to construct an acceptable persona, and yet certain choices encountered resistance. The alchemical persona, it turned out, was deeply embattled from the outset, as contested among practitioners themselves as by outside observers of alchemy's fortunes. Moreover, the persona also turned out not to be something that alchemists could simply assert; the persona had to be culturally and socially meaningful and credible, which meant that alchemists had to reckon with broader social and cultural expectations as well as their own self-image.

In this chapter, I explore the halting emergence of an alchemical persona from the late Middle Ages to the sixteenth century, untangling how some of the initial elements of that persona were reshaped in reaction to public debates about alchemy. As we shall see, the medieval Latin alchemical tradition offered one set of possibilities: the alchemist as scholar, artisan, or even prophet. In the sixteenth century, however, literary sources put forth a different set of possibilities: the alchemist as fool, corrupt merchant, and criminal; in short, as *Betrüger*. This literary persona was incredibly influential, and ultimately alchemists were forced to reckon with it, appropriating the figure of the *Betrüger* as a foil for a new alchemical persona, the expert. In the absence of formal institutions to shape the alchemist's identity, these cultural discourses became all the more important in shaping just who alchemists were and where they fitted in the early modern social and cultural landscape.

Scholar, Prophet, and Artisan

While alchemical authors continued to write about alchemical theory and practice in the early modern period, they also began to comment on the alchemist as a social type, constructing a persona that could, ideally, mediate between their work and their sociocultural positioning. In doing so, early modern alchemists drew in part on medieval tradition, where they found elements of an alchemical persona modeled on other, more familiar, personae: the scholar, the prophet, and the artisan. While none of these three models fitted entirely, collectively they provided a set of resonances that positioned the alchemist as someone with unique access to nature's secrets through mind, spirit, and body.

In the late thirteenth century, one of the most important medieval alchemical texts, pseudo-Geber's *Summa perfectionis* (Sum of perfection), addressed the issue explicitly, constructing an image of the alchemist out of elements of the scholar, artisan, and even prophet. The author of this important text began his treatise with a discussion of "all the impediments by which the practitioner is kept from reaching the true end"—impediments of the body, soul, and "from beyond," that is, from things beyond his control.[6] An alchemist should be physically fit, according to pseudo-Geber, for "if the practitioner does not have his organs in an integral condition, he will not be able to arrive at the end of this work by himself." Any alchemist afflicted, for instance, by blindness, advanced age, or a body "weak or diseased like the bodies of the fevered or leprous, from which the extremities fall off," would certainly be disadvantaged. Even more important than these bodily impediments, however, were the impediments of the soul. Alchemy is not for those with a "simple soul," pseudo-Geber argued, who lack intelligence and whose "head, being full of much smoke, cannot receive the true intent of natural things." Those with a "mobile soul," on the other hand, do not have enough focus and commitment: "passing from one opinion to another, and from one wish to another wish, . . . [they] are so mobile that they can hardly finish the slightest thing that they intend." Pseudo-Geber also dismissed "those deprived of reason, madmen or children," those who condemn alchemy altogether ("whom this science in turn condemns and repels from the most precious end of the work"), as well as those "who are slaves of money," too cheap to invest the necessary funds actually to carry out alchemical laboratory work. Individuals with any of these flaws simply should not attempt alchemy. But even the most model alchemical practitioners, pseudo-Geber acknowledged, might face difficulties: problems "approaching from beyond, from contingent fortunes and mishaps." A sudden shift in fortunes, for example, might force a practitioner "to set this very excellent magistery aside, out of indigence." A practitioner might also be distracted "by the vain cares and temptations of this world." In a nutshell, pseudo-Geber warned, "the impediments coming upon this work are therefore two in general—natural inability and lack of the necessary cost, or of the carrying out of the work."[7]

After describing the array of factors that might hinder alchemical success, pseudo-Geber then went on to describe the ideal alchemist. He (for pseudo-Geber certainly conceived of his alchemist as a man) should be well versed in natural philosophy aided by "natural industry"; experimental practice and study, in other words, must go hand in hand, for "art is aided by ingenuity and likewise ingenuity by art." The alchemist must also be focused, "of constant will in work, so that he does not at one time presume to try one thing, and another time another"; he must also be committed to carry the work out to its conclusion. The alchemist must have "a good temper, and be little given to anger, lest due to the fury of his wrath, he demolish and destroy his work just begun." Interestingly, pseudo-Geber gave great weight to the alchemist's frugality, pointing out that alchemical work could be expensive. The practitioner who spends money too easily on useless things might not have enough to finish the real alchemical work; "these sorts [of alchemists] are doubly in error," pseudo-Geber warned, "both because they dispense their money for useless things and because they lose the noble science which they have sought with squandered wealth. It is not necessary to consume your goods since, if you are not ignorant of the principles of art that we teach you and you understand them rightly, you will arrive at the complete magistery for a small price." Finally, the alchemist must have God's blessing, "since our art is reserved by the divine power, and He who is most glorious, sublime, and filled with all justice and goodness extends it to and withdraws it from whomever He wills."[8]

What sort of persona was this? Not surprisingly, pseudo-Geber drew in large part on the figure of the scholar: learned, focused, committed, but also somewhat detached from the world. Pseudo-Geber, after all, discouraged those "detained by the vain cares and temptations of this world, occupying themselves wholly in every sort of secular business from whom this precious science of ours flees."[9] The alchemist, in other words, should avoid involvement with earthly concerns, presumably things like family, politics, and trade. And yet pseudo-Geber was clearly aware that alchemists needed money. Apparently, then, alchemists (not unlike medieval scholars or clerics) should be either of independent means or reliant on patronage or some sort of benefice.[10] Since the most prominent medieval the alchemists were themselves scholars, the scholar's persona was in some ways a natural fit. And yet it could not account for all of the elements of the alchemist's life and work.

The *Summa perfectionis* also offers hints of the alchemist as prophet, receiving alchemy as *donum dei.* In the fourteenth century, Petrus Bonus of Ferrara would develop this dimension of the alchemical persona, positing an alchemist who was not only a scholar and an artisan, but also a "pious illumine," as Chiara Crisciani has pointed out.[11] In his *Pretiosa margarita novella* (New pearl of great price), this fourteenth-century physician laid out a case for alchemy as a legitimate branch of natural philosophy. At the same time, however, Bonus argued that reason could not unlock all of alchemy's secrets, and that the alchemist must also rely on divine revelation in order to complete the philosophers' stone.[12] "Reasoning is not sufficient for the investigation of the same [i.e., the philosophers' stone]," Bonus wrote, "but one must believe that the end of it can be thus, and so must summon up a profound faith." In fact, Bonus argued, the true ancient alchemists were also prophets: "We can prove incontestably that the ancient philosophers of this art were seers who truly prophesied through this divine art about the manifestation of God in the flesh of man—namely Christ—and his identity with God by means of the inglowing and emmanence of the Holy Ghost."[13] Bonus was not the only alchemical author to take this kind of "religious turn" and emphasize the prophetic dimensions of the alchemical persona. As William Newman has pointed out, a number of other fourteenth-century alchemical texts, including the works of John of Rupescissa, the *Testament of Alchemy* (by pseudo-Lull), and works ascribed to Arnald of Villanova, also stressed "soteriological and eschatological themes." Eventually these claims drew negative attention from theologians. For example, the inquisitor general of Aragon, Nicholas Eymerich, criticized alchemy on precisely this point in his *Directorium inquisitorum* (1376), objecting to alchemists' claims to access the supernatural. With this argument, Eymerich demonstrated, among other things, that the persona of the alchemist-prophet entered into dangerous theological territory.[14]

Part scholar, part prophet, the alchemist of medieval tradition might appear to have been otherworldly. The fact that pseudo-Geber highlighted the importance of the alchemist's physical body, however, serves as a useful reminder that alchemical tradition did not simply emphasize the mind and spirit. Accomplished alchemists, after all, had to carry out demanding work in the laboratory in order to truly understand the art. Like artisans, they grounded their art in the bodily, productive engagement with the materials of nature.[15] Yet adopting the artisan's persona posed a different set of problems, for it carried with it a set of economic, kinship, religious, and political roles that were not entirely appropriate for alchemy.

Unlike scholars, artisans were thoroughly immersed in the world. Artisanal household economies, organized into guilds, were the lifeblood of late medieval and early modern cities. Not only did the producers of shoes, cloth, medicine, and books contribute to the economy and honor of the urban community; they were also one important way in which cities organized social welfare, religious ritual, and, in the many places where guild leaders were represented on town councils, politics. The artisan's work, in other words, was intimately bound up with family, economy, faith, and often politics, tightly woven into the fabric of the city. Alchemists, on the other hand, did not organize their work around guild and workshop, nor were they perceived to be among the many corporate bodies constituting the urban commune. Furthermore, alchemists such as pseudo-Geber specifically advocated avoiding "every sort of secular business," rejecting the artisan's engagement in the world. The artisanal persona, too, apparently had its limits.

In the early sixteenth century, these three models—scholar, prophet, and artisan—persisted as viable ingredients for an emerging alchemical persona even if none could be entirely adopted outright. Scholars, prophets, and artisans themselves had undergone changes, however, evolving from their medieval into more distinctly early modern personae. As Gadi Algazi has shown, for example, celibacy ceased to be the rule among scholars in the late fifteenth century, as they increasingly married and founded households; in doing so, they also transformed what Algazi has called the learned habitus, reorganizing domestic space and cultivating emotional detachment as ways "to live in the world while retaining an otherworldly image."[16] Prophets and artisans, too, acquired new social and cultural meanings in the wake of religious and scientific transformations of the sixteenth century. Eymerich's concerns about some alchemists' claims to divine revelation in the Middle Ages became even more serious as religious conflict and fears of witchcraft took hold in the sixteenth century. Artisans, meanwhile, developed a new self-consciousness grounded in their unique abilities to imitate and know nature and began to rival natural philosophers in their claims to natural knowledge.[17] Even as these models continued to evolve, they endured in their updated forms as possible ingredients for an alchemical persona. A few alchemists still fashioned themselves as prophets in an early modern mold, and even more as scholars.[18] Moreover, as Pamela Smith has shown, the work of the alchemist came to be the paradigmatic example of the artisanal ability to imitate and improve nature that was so central to the development of natural knowledge in the sixteenth and seventeenth centuries.[19] At the same time, new possibilities were beginning to appear, suggested not by alchemists themselves, but by outsiders who looked at alchemy with a much more critical eye. Rather than taking bits and pieces from scholars, prophets, or artisans, these critics constructed the alchemist's persona of different ingredients altogether: the fool, the corrupt merchant, and the criminal.

"O What Rotten Promises, What Stupid Credulity!" Humanist Satire

In order to fully succeed, a persona must be recognizable, as well as socially and culturally credible. Alchemists could not simply assert their new persona, in other words, and assume that it would prevail; the persona also had to find broad social and cultural acceptance and recognition. In this regard, early modern alchemists ran into difficulties, because they were not the only voices proposing an alchemical persona. In fact, an entirely different alchemical persona had found firm footing in the Holy Roman Empire by the early sixteenth century, and humanists and urban elites increasingly found it far more plausible than the persona that alchemists were constructing for themselves. This newer, competing persona was the alchemist as *Betrüger*, and it turned up in the works of prominent humanists such as Francesco Petrarch, Sebastian Brant, and Desiderius Erasmus. Some medieval alchemical authors had already sketched out a kind of false alchemist: the sophist, whose understanding of alchemy was as shallow as his transmutations, altering only the surface qualities of metals, not their true nature. The figure of the sophist, however, was grounded in a critique of knowledge, understanding, and skill. Humanists, on the other hand, only rarely mentioned philosophical debates about alchemy. Instead, they launched their attack against something new: the intention, character, and morality of the alchemist, lambasting the *Betrüger* as everything from greedy, addicted fools to pernicious, predatory liars. In a sense, the humanist critique took the older argument that sophistical transmutations were merely superficial and personified it. Now the alchemist-*Betrüger* himself was as superficial as his gold: an alchemist in appearance only, not in his true nature. This fundamental untruth led only to further deceptions and moral degradation. From this critical perspective, alchemists offered rich material for social satire alongside the sturdy beggars, unchaste housewives, incompetent physicians, and corrupt clergymen who also came up for mockery in early modern satirical texts.[20]

As these literary authors sketched out the figure of the alchemist for their readers, this fictionalized character seemed to come to life in more and more detail. The persona of the alchemist-*Betrüger* could be a pitiful fool, transformed by obsession with the art into an unrecognizable character; a greedy, corrupt peddler, driven by the desire for money to take advantage of naive dupes; or a base criminal who threatened to ruin society. Assembled out of the literary tropes of obsession and deception, this persona was certainly negative, but this did not make it any less productive. Paradoxically, by establishing a counterpoint against which alchemists could assert a new, more powerful persona as expert, the *Betrüger* played a crucial role in the development of alchemy in the early modern period.

The humanist attention to the alchemist was part of a much broader trend, dating to the fourteenth century, of vernacularization and laicization in European alchemy.[21] When alchemical ideas first appeared in Europe in the twelfth century, they were specialized knowledge for scholars. Whether translations of older Greek, Arabic, and Hebrew texts or newer texts written by Europeans, twelfth- and thirteenth-century alchemical texts were written in Latin for a limited audience. As William Newman has demonstrated, alchemy engaged these schoolmen because it lay at the heart of a debate about the power of art, or technology, in relationship to nature. In part because it had wide-ranging implications beyond alchemy, from the visual arts to demonology, this debate around alchemy continued to fuel intellectual debates well into the seventeenth century.[22] In the fourteenth century, alchemy began to appear in a wider variety of genres and languages; at the same time, literary authors also began to comment on alchemy. Historians of medieval alchemy have argued that both the vernacularization of alchemical texts and the rise of "literary witnesses" to alchemy, as Michela Pereira has called them, in the fourteenth and fifteenth centuries are evidence of the diffusion of alchemical ideas beyond the fairly narrow world of university-educated scholars.[23]

This late medieval shift in genre and audience was absolutely crucial for debates about alchemy, for it turned what had been a largely philosophical and theological debate about alchemy into one about the alchemists as a social type. These "literary witnesses" increasingly wrote about alchemists as the kind of person one might encounter in the marketplace, or even, as in Dante's *Inferno,* in the final circle of hell.[24] Although a number of earlier literary texts briefly mentioned alchemy, the first sustained literary discussion of the alchemist appeared in Petrarch's moral treatise *De remediis utriusque fortune* (Remedies for fortune fair and foul, 1358–61), which considered the dangers of both prosperity and adversity. Shortly thereafter, Chaucer also devoted an entire section of his *Canterbury Tales,* "The Canon's Yeoman's Tale," to alchemy, creating an image of the alchemist that would prove to be influential in England into the seventeenth century.[25] In the sixteenth-century Holy Roman Empire, however, Chaucer was insignificant in comparison to Petrarch; this was particularly true after the appearance in 1532 of an extraordinarily successful illustrated German translation of *De remediis,* entitled *Von der Artzney bayder Glück, des guten und widerwertigen,* made Petrarch's text accessible to German vernacular readers who may not have known the Latin original.[26] Humanist authors in northern Europe later picked up Petrarch's thread, including the figure of the alchemist in their own moral and satirical treatises. Alchemists appeared, for example, in Sebastian Brant's (1457–1521) astonishingly successful satire *Narrenschiff* (Ship of fools, 1494), while Desiderius Erasmus (ca. 1466–1536) devoted two of his colloquies, *Alcumistica* and *Πτωχολοια* (Beggar talk), first published in 1524, to the subject.[27] Heinrich Cornelius Agrippa of Nettesheim (1486–1535) addressed the subject as well in his treatise on the uselessness of art and knowledge, *De incertitudine et vanitate scientiarum* (On the vanity and uncertainty of arts and sciences, written in 1526, published in 1530), as did Johannes Clajus (1535–92), somewhat later, but more directly, in his cleverly titled 1586 *Altkumistica, das ist: Die Kunst, aus Mist durch seine Wirckung, Gold zu machen: Wider die betrieglichen Alchimisten, und ungeschickte vermeinte Theophrasisten* (*Altkumistica;* or, The art of making gold out of dung: Against the fraudulent alchemists and unskilled Theophrastians).[28] Together, these and other authors constructed a significant critical discourse around alchemy, drawing on the fool, merchant and criminal to make the alchemist into a potent symbol of human foibles. A closer examination of these texts, then, is crucial for understanding the challenges that confronted alchemists as they strove to create their persona.

Von der Artzney bayder Glück, the 1532 German edition of Petrarch's *De remediis utriusque fortune,* established the image of the alchemist as fool, the pitiful victim of a hopeless obsession with alchemical success. In this text, Petrarch offers readers a dialogue between Reason and the four human passions of Joy, Hope or Desire, Sorrow, and Fear; this conversation was to serve as a model of how every thoughtful human should best confront the "fair and the foul" that Fortune has to offer. Near the end of book 1, which deals with remedies for prosperity, *Vernunfft* (Reason) and *Freud* (Joy)[29] discuss the good things one might hope for in the future, including "expecting an inheritance," "hope for fame after death," "the promises of fortune tellers," "hope for life eternal," and "alchemy." As Joy proclaims in a series of terse statements that she hopes that alchemy will bring her "good results," "success," "gold," "great things," and "riches,"

Reason responds to each declaration with a much longer counterexample or argument. By modeling their own decisions on this kind of debate, presumably, Petrarch's readers would have learned how to make a well-considered decision about anything, in this instance about undertaking alchemy.

In countering Joy's optimistic predictions, Reason lays out a devastating case against alchemy. Neither Joy nor anyone else has witnessed alchemy (which Petrarch defined here narrowly as transmutation), Reason declares, and "all reports that it happened to some are fabricated by people who have found it expedient to believe so."[30] Believing alchemical success to be unattainable, Reason naturally predicts that the outcome of Joy's alchemical endeavors will be disastrous, leading to a catastrophic mix of madness, obsession, poverty, and even degeneracy and blindness:

> Goaded by greed and mad impulse, you deem to be true what you hope for, and false what you have before your eyes. You must know some who, as a rule, are reasonable, but in this respect utterly mad, and some very wealthy people, wholly consumed by this foolishness—and while they try to become wealthier lusting after filthy profits [*schöden gewinn*], they waste what is properly theirs and spend all their riches on useless things and, eventually, are deprived even of necessities. Others forsake their *bürgerlich* and courtly mores and live full of sadness and anxiety, unable to think of anything save bellows, tongs, and coals; [they] live with nothing other than their fantasy and become some sort of wild creatures. In the end, many lose first intellectual vision [*augen des gemuts*] and then the use of their physical eyes [*ubung die leyplichen augen*] to boot.[31]

In this passage, Petrarch dismisses quickly the question of whether it is possible for alchemists to transmute metals, focusing instead on the way in which alchemy transmutes alchemists themselves. Reason concludes that alchemy ultimately will lead to "needless worries, a foolish heart, a disfigured and grimy mouth, blinded eyes, and painful poverty, and worst of all, the name of a fraud and a dissolute rascal [*den namen eins Gaucklers und Loterbubens*] and a life spent in the darkness of the night and the secret lurking holes of thieves."[32] Destroyed economically, emotionally, physically, and socially, in Petrarch's eyes the alchemist is a pitiful figure indeed.

The message in *Von der Artzney bayder Glück* was underscored by hundreds of woodcuts—one for each chapter—that were certainly responsible

Das Erſt bůch/von Artzney
Vonn Alchemey/Das
CXI.Capitel. Freud.

Ch hoff der Alchemey glückſäligen außgãg. Der.
Es iſt ein wunder/von wannen du das hoffeſt/das dir doch nye
warhafftig iſt begegnet/auch ſunſt niemant/vnd ob wol das ge-
ſchray were/etlichen zůgeſtanden ſein/iſt erdichtet/von denenn ſo
jnen ſollichs zů gelauben nutz machen. Freud. Ich hoff die Alchemey werd recht

Figure 5. Woodcut depicting alchemical folly in Francesco Petrarca, *Von der Artzney bayder Glück, des guten und widerwertigen* (Augsburg: Heynrich Steyner, 1532) (German translation of the original Latin *De remediis utriusque fortunae libri duo*, 1360/66). (Typ 520.32.683 F (A), Department of Printing and Graphic Arts, Houghton Library, Harvard College Library.)

in large part for the popularity of the German edition.[33] The illustration accompanying Petrarch's chapter on alchemy gave visual form both to the critique that Reason lays out in the text and to the persona of the alchemist (fig. 5). The poverty of the alchemist and his assistant is apparent in their clothes, which are covered with holes and tears. The alchemist stands at the furnace, peering into a vessel through a pair of glasses perched precariously on his nose, recalling Petrarch's warning that alchemy would produce nothing but "a disfigured and grimy mouth, blinded eyes, and painful poverty."[34] The assistant, meanwhile, stands idly by, scratching his shaggy head in confusion. This is not the image of a successful alchemical enterprise; the chaos of the laboratory evokes the degeneracy that Petrarch suggested would befall anyone who became obsessed with alchemy.

The power of alchemy to transform body and soul, even if it could not transmute metals, appeared as a theme in a number of other early modern texts as well. Petrarch probably inspired Sebastian Brant to take up the theme in both text and image in his 1494 *Narrenschiff.* Brant certainly knew well the commentary on alchemy in *De remediis,* for he was closely involved in the production of the 1532 German translation; Brant contributed a prefatory poem, and the printer's preface (written after Brant's death) explained that the book contained "many elegant and wonderfully delightful images, which have been provided for each chapter according to the design of the late learned Doctor Sebastian Brant."[35] In the *Narrenschiff,* Brant described 112 types of fools and follies, ranging from the collector of "useless books" to the beggar, noise in church, foolish medicine, quarreling and going to court, and becoming a priest.[36] Each fool or folly earned a chapter, including a three-line motto, a woodcut, and a verse description of the vice at hand. As John Van Cleve has pointed out, a central concern of the *Narrenschiff* is that "a preoccupation with money and property makes happiness in this world possible, but not inevitable, and salvation in the world to come impossible."[37] Alchemy is but one of a myriad of worldly distractions that ultimately can jeopardize the soul. In his discussion of alchemy, Brant echoes Petrarch's remarks on alchemy's transformative power, contrasting the real transformation of the alchemist with his failure to transmute metals. However well intentioned, Brant argues, alchemists will most likely reap little gain but pay a high price for their practice, for alchemy's addictiveness destroys its practitioner and everything else. Not only does the alchemist who spends all his time peering[38] into the crucible eventually drive others out of his household, but through his obsession with alchemy, the alchemist transforms himself utterly beyond recognition: "He who previously sat gentle and prosy, prods his material in the monkey's glass, burning it down to a powder, until he no longer knows even himself." Pecuniary peril soon follows: "Many have ruined themselves, but precious few have earned financial gain." For Brant, as for Petrarch, alchemy transforms everything but the metal, as families leave, patrimonies disappear, and alchemists themselves are nearly deformed by their foolish obsession.[39]

This image of the alchemist-fool as an object of pity continued to appear in the sixteenth century.[40] Erasmus too described alchemy as almost bewitching—"a disorder so intoxicating, once it strikes a man, that it beguiles even the learned and prudent."[41] In his colloquy on *Alcumistica*

(1524), however, he turned the tragedy of alchemical obsession into an object of humor, foreshadowing works such as Ben Jonson's play *The Alchemist* (1616), which used alchemy to mock human foibles. In this colloquy, Erasmus described a ridiculous set of alchemical scams by which a priest "who understood the 'art' [of alchemy] about as well as an ass does" tricked a "much esteemed gentleman" named Balbinus out of his fortune. In this story, the priest-alchemist relied entirely on his silver tongue to convince Balbinus to invest money in alchemical projects. All the priest had to do was reveal that he had a lifetime of experience with alchemy and praise Balbinus's own learning; then, "after the sly old rascal, by this kind of talk, had dispelled suspicion of fraud and had convinced Balbinus of his perfect understanding," Balbinus easily agreed to fund further alchemical work. "Money is counted out then and there for the operator to buy pots, glasses, charcoal, and other equipment needed for the laboratory," Balbinus's friend Lalus explains. "This money our alchemist squanders enjoyably on whores, dice, and drink." The priest-alchemist stalled again and again, offering a series of excuses about bad charcoal, improper heating of glass vessels, offerings to the Virgin Mary, bribes to officials, and the priest's own sinfulness in order to explain why he needed more money and why the alchemical process was not yet finished. Poor Balbinus, meanwhile, "devoted all his time to calculations . . . figuring how much profit, if one ounce yielded fifteen, would be made from two thousand ounces."[42]

Eventually, "after he had made a fool of the man for quite a while by tricks of this sort, and fleeced him of no mean sum of money," an old acquaintance of the priest-alchemist denounced him as a scam artist. Balbinus, mortified by his own gullibility, simply "gave him travel money, imploring him by everything sacred not to blab about what had happened," and sent him on his way. In Erasmus's telling, Balbinus has only himself to blame for this debacle. His friends Lalus and Philecous, who relate this tale in the colloquy, are not particularly disturbed either; they are simply puzzled and amused by Balbinus's weakness, attributing it to the "notorious disease" of alchemy. Indeed, poor Balbinus's disease seems to be chronic, for Lalus explains, "Balbinus, often as he's been taken in by its practitioners, nevertheless allowed himself to be marvelously cheated a little while ago." In the end, Philecous concludes, "I might feel sorry for Balbinus if he himself didn't enjoy being gulled." For Erasmus, little harm is done, save for a diminished fortune and perhaps a bit of social disgrace.[43]

Over the course of the sixteenth century, however, this humorous image of the alchemist as the maker of fools was slowly overshadowed by a more pernicious figure: the predatory alchemist as deceiver, or *Betrüger*.

This newly prominent alchemical *Betrüger* was just as much a literary creation as his intoxicated forbears, but this newer construction also resonated with a different intellectual climate. For Petrarch, the alchemist served as a moral exemplar, a warning against self-delusion and a vehicle for exploring how individuals should weigh the unlikely promise of future gain against its high costs in the present. Later humanists like Erasmus, however, increasingly directed their satirical eye toward the costs of alchemy for society at large; here the alchemist's deception of himself is turned outward, deceiving others now instead. With this shift, it made sense that critics would turn to a salient metaphor for human interaction in the early modern period—commerce—and model this new alchemical *Betrüger* in part on the merchant.

This connection between alchemists and fraud, like that between the alchemist and the fool, has its roots in Petrarch's *De remediis*. Although he focused mainly on alchemy's seductive power, Petrarch attributed that power to the lies that he believed lay at its core. Alchemists claim that their art is true and that transmutation is possible, even if they themselves have not yet achieved it. And yet, Petrarch insists, these claims are untrue, and so anyone who propagates them is a liar; whether or not Petrarch sees alchemists as lying to themselves or to others, however, is unclear. On the one hand, he hints that they are con men—knowing liars who self-consciously spread falsehoods in order to take advantage of those gullible enough to believe them. "He who promises you gold will unexpectedly run away with the results of your work," Reason explains to Joy; "I tell you that the art you refer to is nothing else but the art of lying and cheating [*betriegen*]." And yet Petrarch's alchemists mostly seem to lie to themselves, preferring always to listen alchemy's false "hope and imaginings," as Chaucer called it, rather than admit failure.[44] Reason's partner in this dialogue, Joy, represents the point of view of the ever-optimistic alchemist. "I am close to the end of my efforts," Joy earnestly insists. Reason, however, tries to dispel this delusion, insisting, "You have easily congealed mercury (or quicksilver) or accomplished some other fantasy, but you are now as far as ever from the end of your undertaking. You will forever lack some essential ingredient, but you will never lack falsehood [*triegerey*]."[45] In the end, it did not matter to Petrarch whether alchemists deluded themselves or others; either way, the art was so thoroughly corrupted by falsehood that it could only lead to ruin.

Sebastian Brant elaborated much more fully on this theme and linked it explicitly to commerce in his 1494 *Narrenschiff.* Alchemists appear in Chapter 102, "Vom Falsch und Beschiß" (Of falsity and deception), alongside a crowd of people dealing in various types of deceit: those who sell lame horses as noble steeds, employ false weights and measures, debase coins, give false counsel, claim false piety, sell "mouse shit [*Mausdreck*] for pepper corns," hawk dyed furs and rotten herring, and adulterate wine with "saltpeter, sulfur, bones of the dead, potash, milk, mustard, impure herbs." Deception was everywhere, Brant declared, bringing spiritual corruption, death, and ruin; moreover, it flourished because of greed and the drive for profit. "False love, false counsel, false friend, false money," he lamented. "The entire world is full of fraudulence! Fraternal love is dead and blind, everyone thinks only of fraud and illusion; one wants only to earn without loss, even if hundreds are ruined in the process. Honor is nowhere to be seen anymore, [and] a man would give up salvation, if only to sell off a single thing; God's mercy to whoever dies in the meantime!"[46] As John Van Cleve has noted, this "withering evaluation equates commerce in all forms with deception."[47]

What role did Brant imagine alchemists playing in this general atmosphere of wordly corruption and deceit? Brant drew on the increasingly important early modern figure of the merchant in creating the alchemist-*Betrüger*; and for him, the merchant was a deeply flawed figure. In part, he argued, alchemists contributed to a general problem plaguing the marketplace: the price of commercial goods, in Brant's eyes, had come unhinged from real value, as merchants of all types relied on deception to convince buyers to pay good money for shoddy products. "No shopkeeper's goods have fixed value," Brant objected; "everyone seeks to trade in deceit, just to unload the goods, even though the quality is terrible."[48] Alchemy, for Brant, was as an example par excellence of this manipulation of quality for profit. Just as peddlers relied on tricks and superficial alterations (such as painting furs) to make their goods seem to be something that they were not, so too did alchemists "perform crude sleights of hand and deceptions" (Sie gaukeln und betrügen grob). Brant laid bare the deceptions that vendors and alchemists purportedly used to trick would-be buyers. He claimed, for instance, that the *Betrüger* concealed precious metal in a hollow rod before they began a supposed transmutation, plugging the bottom of the rod with a bit of wax. As they stirred the molten contents of a crucible during the supposed transmutation, the wax melted, releasing the gold or silver into the mix and thus ensuring that the resulting metal could withstand a goldsmith's assay. This trick, which Brant was not the first to describe, would become a commonplace in descriptions of alchemical legerdemain.[49] By describing such tricks alongside other kinds of commercial deceit, Brant linked alchemy to all kinds of corrupt commercial practices, suggesting that the alchemist's persona was connected to the worst hucksters of the marketplace.

And yet, in Brant's eyes, alchemical tricks were unique. Whereas the other sorts of peddlers he describes were trying to sell false *products*, Brant's phony alchemists were trying to sell false *knowledge*. After all, the gold they produced with hollow rods really *was* gold, it just was not produced through an actual transmutation of quicksilver, as the alchemists led their customers to believe; instead, the crucible "bears gold and silver / which earlier was put in the rod [*stäcklin*]."[50] The false transmutation was not meant to be an end in itself, but rather a demonstration of the quality of the real product for sale, the knowledge of how to transmute metals. The alchemist's parallel was not the wine seller who hawked casks of mustard and milk as if they were real wine, but rather a wine seller who offered customers a sample taste of good wine from one cask, but then surreptitiously switched casks so that the customer took home the bad wine instead. The false merchandise that the fraud was designed to conceal, then, was not the alchemical gold or silver, but the process or recipe for creating it, which was the real item for sale in this false transaction. Brant imagined alchemists as false peddlers of knowledge, though not of false precious metal, in his marketplace of deception.

Brant left readers with little doubt that some alchemists belonged in his ship of fools. What he did not say, however, was that *all* alchemists belonged on board. Brant still left open the possibility that some might succeed, and here he set up a differentiation among the deliberately deceptive alchemist (the *Betrüger*), the incompetent fool (the sophist), and the real alchemist, who succeeds honestly. These distinctions, reflected in Brueghel's threefold image of the alchemist, would persist well into the seventeenth century. The woodcut that accompanied the *Narrenschiff*'s chapter on falsity heightened the ambiguity of his condemnation (fig. 6).[51] The vintner on the right and the alchemist on the left, both in fool's caps, clearly represent the "falsity and deception" that is the subject of the chapter, as the motto below the woodcut in the original German edition makes doubly clear: "One easily sees in alchemy, as well as in the doctoring of wine, the lies and fraud there are on earth" (Man spurt wohl in der Alchemei, und in des Weines Arzenei, welch Lug und Trug auf Erden sei). The alchemist-fool on the left of the image, his robes ripped in places, pokes at a vessel in the fire (or perhaps stirs it with a hollow rod?), recalling the image accompanying Petrarch's *Von der Artzney bayder Glück*. The crucibles scattered on the floor hint at a disorderly laboratory. There is a third figure in the image, however, who does not wear the fool's cap and whose

FOLIO

De falſariis & fraudulentia.

Alchimia docet fallax: corruptio vini &:

Quas fraudes hominũ perfidus orbis agat.

Omnibus in rebus fraudis falſiq́ꝫ/doliq́ꝫ:

Cuncta ſcatent vitio: tuta nec vlla fides.

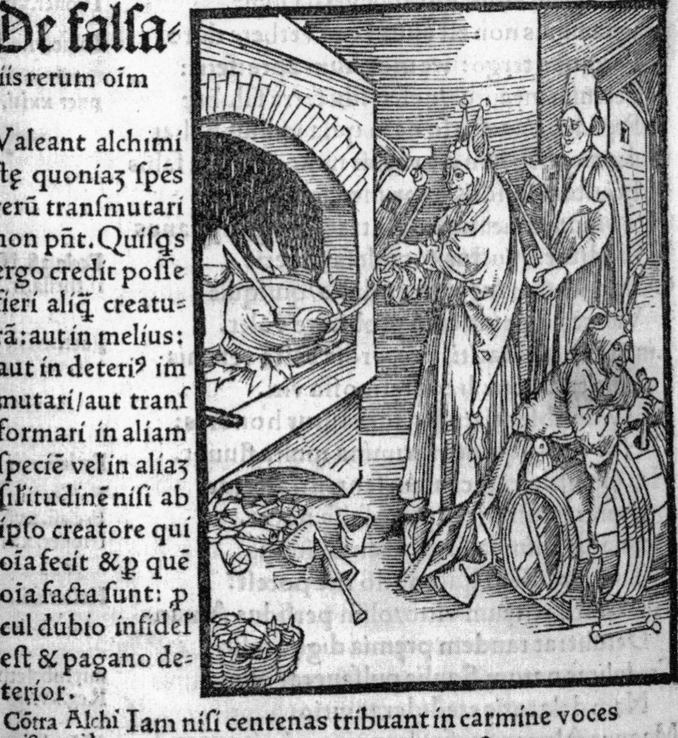

De falſa-riis rerum oim

Valeant alchimiſtę quoniaꝫ ſpẽs rerũ tranſmutari non pñt. Quiſq́ꝫ ergo credit poſſe fieri aliq̃ creaturã: aut in melius: aut in deteri⁹ immutari/aut tranſformari in aliam ſpeciẽ vel in aliaꝫ ſil'itudinẽ niſi ab ipſo creatore qui oĩa fecit & ꝑ quẽ oĩa facta ſunt: ꝑcul dubio infidel' eſt & pagano deterior.

Cõtra Alchimiſtas. vide tex. in. c. epi circa ſi. xxvi q. v.

Iam niſi centenas tribuant in carmine voces

Caſtalię: totidem linguas: vireſq́ꝫ potentes

Dictandi: nequeam falſos componere ſtultos.

Copia magna quidem eſt illorũ: magna caterua

Figure 6. Woodcut accompanying the chapter "De falsariis" in Sebastian Brant, *Stultifera navis* (Basel: Johann Bergmann de Olpe, 1498). (Brown University Library.)

robes seem less worn. His dress and stature, in fact, suggest that he is not the foolish alchemist's assistant, as in *Von der Artzney bayder Glück,* but rather a man of higher station; perhaps he is meant to represent a more scholarly or a more honest kind of alchemy, as opposed to Brant's *Betrüger*-fools. And yet, insofar as the second alchemist does not seem completely separate from the *Betrüger*-fool, one wonders about their relationship. Is the second alchemist looking on, luring the fool into what he knows is a hopeless pursuit?[52] Or is he, too, attracted to the attempt to transmute metals, a suggestion that the folly of alchemical transmutation could draw in anyone? In visually articulating this tension with two separate figures on the left, the woodcut artist adds a layer of complexity and ambiguity to the text, opening a possibility that not all alchemists were *Betrüger.* Brueghel, of course, would heighten this ambiguity a half century later in his own alchemist engraving.

By including alchemists among those who sell false wares, Brant brought alchemy into the arena of trade and commerce and thus fleshed out a dimension of the *Betrüger* that was not yet present in Petrarch. Brant positioned alchemists as yet another kind of false merchant, linking alchemy not only to an older view of money and greed as a sinful attachment to the earthly realm, but also to a much newer discussion about the uncertain relationship between value and price in the nascent commercial economy of early modern Europe.[53] If price was pegged to the *perception* of worth—whether in social, cultural, intellectual, or economic terms—then deception was particularly troubling, because it revealed the fragility of the model. Erasmus took up this theme in a second colloquy on alchemy, *Πτωχολοια* (Beggar talk) which dealt with an impostor who poses alternatively as a beggar and an alchemist in order to dupe people out of their money. In this colloquy, a former beggar named Misoponus ("hater of work") confesses to another beggar how he managed to play both roles. As a beggar, he had been covered with sores; however, as he explains, "all that decoration of mine I put on with paints, turpentine, sulfur, resin, birdlime, linen cloth, and blood. When I felt like doing so, I took off what I had put on." When Misoponus's friend charges, "Impostor! There was nothing more wretched-looking than you were," Misoponus explains simply, "My need compelled it at the time; though Fortune too sometimes changes her skin." Misoponus then describes how he turned those same imitative skills to alchemy, performing false transmutations (with hollow coals filled with silver) in order to convince people to buy a supposedly transmutative powder for "a large sum." His friend asks, "Is your profit from this profession enough to support you?" "Oh yes," Misoponus answers;

"in fine style, too. Hereafter, if you're smart, you'll give up this wretchedness and join my order."[54]

By describing alchemical *Betrüger* as an "order," Erasmus hinted at the existence of an alchemical underground, adding the alchemist-*Betrüger* to the colorful rogues' gallery that had come to fascinate Europeans in the sixteenth century. Just as Erasmus and his contemporaries collectively detailed the imagined tricks, language, and motivations of false alchemists, so too did a burgeoning literature of roguery introduce readers to the organization and culture of an imagined society of beggars, vagabonds, and criminals at the margins of European society.[55] These texts reflect a variety of attitudes to this imagined subculture, which could simultaneously spark curiosity, delight, fear, and disdain. By suggesting that Misoponus belonged to an existing "order" of fraudulent alchemists, Erasmus alluded to the possibility that alchemical fraud was not just an isolated problem; perhaps criminal alchemists were just as pervasive and organized as Europeans imagined other vagabonds and beggars to be. In fact, he insinuated, criminals and alchemists might be one and the same, as someone practiced at fraudulent begging could just as easily turn those skills to impersonating an alchemist; a life of crime, in other words, could easily lead to false alchemy. The artist Barthel Beham (1502–40), on the other hand, worried about precisely the opposite. His woodcut *Twelve Vagrants* (ca. 1524) depicted twelve different kinds of beggars, all with short poems describing how they fell into poverty. As the very first beggar explains, "the fact that I stand here naked, and must go begging throughout the land, was brought about by alchemy." Rather than crime leading to alchemy, Beham feared that alchemy could lead to poverty, which was increasingly synonymous with crime in the sixteenth century.[56] The humanist, alchemist, and natural magician Heinrich Cornelius Agrippa expressed the same sentiment in his 1526 *De incertitudine et vanitate scientiarum atque artium declamatio* (On the vanity and uncertainty of arts and sciences): the financial and social degeneration that inevitably followed most alchemists could easily force them into moral collapse. "In total poverty they are compelled to beg," he claimed, "and in such a state of misfortune, rather than favor and mercy, they receive contempt and ridicule. Their plight leads them on to wicked arts [*malae artes*], such as counterfeiting coins and other falsifications."[57] By linking alchemy to crime via poverty, even hinting at an "order" of fraudulent alchemists, Erasmus, Beham, and Agrippa were able to condemn alchemical fraud while also exploiting readers' growing curiosity about this hidden world that seemed to surround them.[58]

Whether they earned a living through false begging or false alchemy, the problem with Misoponus and other "haters of work" was that they did so through deception rather than the honest sweat of their brow. Agrippa too drew on this as a reason to condemn alchemists, reminding his readers of God's commandment to work: "Alchemists are of all men the most wicked [*perversissimos*], since indeed God commanded: you shall eat bread from the sweat of your brow, and elsewhere he said to the prophet, you will be nourished by the labor of your own hands, and so you are blessed and you will be well. These [alchemists] despise divine commandments and want the promised blessedness without work, or, as is said, to make mountains of gold only with women's work and childsplay."[59] Decades later, the prolific philologist, educator, theologian, and pastor Johannes Clajus took up this moral critique of alchemists as too lazy to do proper work in his *Altkumistica*.[60] Clajus urged his readers to shun the get-rich-quick schemes of the alchemist and to embrace instead honest *Altkumist*, or "old-cow-dung," if they wanted to make gold. Simply feeding cows and oxen would produce ample dung, Clajus pointed out, to fertilize fields and grow grain, which could in turn feed cows, pigs, sheep, geese, and hens to produce a cornucopia of other agricultural products. From these animals and crops, one could produce leather shoes, wool and flax clothing, meat, beer, wine, flowers, fruit, and even feather pens to write poetry or prognostications. Clajus concluded,

> All of this comes from dung,
> When the land is fertilized with it,
> And it follows that *Altkumisterey*
> Is the proper art of making gold.
> Because what is reported here is that,
> When all of this is made into money [i.e., sold],
> One can get gold for it
> And have forged whatever one wants. . . .
> Thus I praise Altkuhmisterey,
> In which there is no fraud [*Finantzerey*],[61]
> With God it is sure and certain,
> [It] bears gold from dung, *Probatum est*."[62]

For Clajus, it was simply self-evident that alchemy was associated with fraud and "false trade"; he merely had to state the connection, already forged by Petrarch and his followers, as a foil for his praise of the rural economy.

By the time Clajus published *Altkumistica* in 1586, many different rhetorical threads had come together in the persona of the *Betrüger*. This sixteenth-century alchemist—part fool, part corrupt merchant, and part criminal or even loafer—was a distorted reflection of the persona medieval alchemists had tried to create for themselves. Nevertheless, the *Betrüger* took hold in the European imagination because it made alchemy meaningful. Constructed out of broader cultural discourses about moral weakness, the marketplace, and even a criminal underground living at the margins of society, this persona was widely recognizable, even socially and culturally meaningful, in a way that the alchemist's medieval persona never quite was. Above all, alchemy became a ready symbol of deceit, whether self-delusion or the deception of others. In this sense, the *Betrüger* was a personification of and answer to the medieval scholarly debate about whether alchemy could ever truly surpass nature or only hope to imitate it superficially. Although some, like Brant, left open a tiny possibility for true alchemical transmutation, the loudest answer, from Petrarch to Clajus, was that alchemists could only *appear* to be successful through their stunning powers of imitation and deception.

"Fie, You Fraudulent Goldmaker!"

This new persona, the alchemist-*Betrüger*, did not go unnoticed by alchemical authors. As influential humanists created a negative image of the alchemist designed to serve their larger social critiques, they threatened to undermine the persona alchemists were creating for themselves. Even worse, from the perspective of alchemy's supporters, was the fact that humanists like Brant and Erasmus attracted readers who might otherwise know little about alchemy; they may have even reached the illiterate, who could not read but could view the woodcuts accompanying texts like Brant's *Narrenschiff*. Whether humanist scholars, young Latin students, or urban elites, many of these readers and viewers would have been unlikely to pick up a Latin treatise on alchemical theory that treated the subject seriously, but certainly they could enjoy reading witty verses deriding alchemists or contemplating the complicated visual arguments of the accompanying woodcuts. As a result, early modern Europeans could easily develop a one-sided view of alchemy grounded only in the humanist critique, rather than the alchemists' own discussions of their art. As the sixteenth century neared its end, alchemists such as Leonhard Thurniesser and Andreas Libavius, as well as the authors of a host of less well-known vernacular texts, could no longer ignore the *Betrüger*. Confronting this rival persona, in fact, became just as compelling a reason for alchemists to take up their pens as articulating new ideas about alchemical theory and practice.

In acknowledging the connection between alchemy and deceit taking root in the minds of German readers, alchemical authors did not simply concede defeat. Rather, they sought to take control of this discussion, repositioning themselves in the European imagination as important allies in the fight against alchemical *Betrug,* rather than perpetrators of it. Alchemical texts such as the *Alchimyspiegel* (Mirror of alchemy, 1597), Leonhard Thurneisser's *Magna Alchimia* (1583), and Heinrich Khunrath's "Treuhertzige Wahrnungs-Vermahnung eines Getreuen Liebhabers der Warheit an alle wahre Liebhaber der Naturgemässen Alchymiae *Transmutatoriae;* daß wegen der Bübischen Handgriffe der betriegerischen Arg-Chymisten gute Auffacht vonnöthen" (Heartfelt warning and admonition from a faithful devotee of the truth to all true enthusiasts of the natural transmutational alchemy, which necessitates close attention because of the roguish tricks of the wicked *chymists*) all explained how to recognize the tricks of false alchemists as a humble warning to princes and other wealthy consumers who might otherwise fall prey to alchemical treachery. In place of *Betrug,* of course, these authors offered up their own authentic alchemical know-how. Rather than giving in to increasingly negative representations of alchemists in print, therefore, these authors found a way to turn them to their own advantage, using the figure of the *Betrüger* as a foil for their own social and cultural agendas. In the process, alchemical authors in the late sixteenth century cultivated yet another persona that promised to find broad cultural and social acceptance: the alchemist-expert who could navigate the alchemical marketplace that Brant described and distinguish real alchemists from frauds.[63] A crucial starting place for the authors of all of these books was the simple point that all alchemists were not alike. In countless texts from the end of the sixteenth century, both in the vernacular and in Latin, alchemical authors sought to differentiate among different types of alchemists. They argued that there were honest alchemists as well as impostors, and that even if a few criminals falsely posed as alchemists to perpetrate their scams, surely this should not condemn the art as a whole.[64] One vernacular text that appeared in 1597, *Alchimyspiegel,* exemplifies this strategy, as its long title makes clear: *Alchimyspiegel: oder Kurtz entworffene Practick, der gantzen Chimischen Kunst: neben Anzeig, welche darzu tüglich seyen, oder nit: Wie der Alten mit seltzamen verdunckelten Reden und Wörtern hievon beschrieben Bücher zu verstehen: Und darinnen sonderlich der falschen Alchimisten Betrug entdecket wird. Alles in zweyen lustigen Gesprächen verfasset: Und das erste vor diesem aufs dem Arabischen von Roberto Castrensi in Latein, Nun aber sampt dem andern so newlich Lateinische beschriben worden, in unser Teutsche Sprach ubergesetzt, Durch Teophilum Caesarem August* (Mirror of alchemy; or, the practice [or trick] of the entire chymical art, briefly sketched, with information about what is virtuous about it and what is not, how to understand the books of the ancients, written with obscure language and words, and in which the deceit of the false alchemists will be especially revealed, all written in two delightful dialogues, the first of these [translated] from the Arabic into Latin by Robertus Castrensis, the other recently written in Latin, [both] translated into our German tongue by Theophilus Caesarus August).[65] Generally speaking, Caesar himself was supportive of alchemy. As he explained in the preface to the *Alchimyspiegel*, in fact, he translated these two dialogues from Latin into German precisely because he wished to demonstrate alchemy's value as the foundation of numerous other arts, from minting coins to making guns. The first entries in the book, one of the first texts to be translated from Arabic into Latin in the twelfth century and transcribed and printed innumerable times thereafter, purported to be a seventh-century dialogue about the art of alchemy between a Christian monk named Morienus and an Arabian prince named Khalid ibn Yazid. It was the second dialogue, however, that served Caesar's desire to condemn, as he put it, those who "deceitfully and deliberately dare to lure in [*anzubringen*] inquisitive people, whom they then detain with great false hopes long enough until they empty their pockets; [the alchemists] see their profits, and then finally leave behind manure and dung for the investor in place of his gold."[66] This anonymous text, a dialogue between two characters named Theophrastus and Chrysophilus ("that is, the Gold-Grubber [*Gold-geitziger*] or Gold-Lover"), originally appeared in print in a 1561 compilation of Latin alchemical texts entitled *Verae Alchemiae artisque metallicae*.[67] Caesar's German title for this dialogue emphasized the issue of fraud, promising that "the false and despicable [*Leckerisch*] inventions [*gedicht*] and deceits will be detected."[68]

The dialogue between Chrysophilus and Theophrastus begins and ends with the issue of *Betrug*. As Chrysophilus greets his friend Theophrastus, he confesses "not only how much money but also (and this is even more lamentable) how much time I have spent futilely in the hopes of learning the art that the learned men call *chymica* . . . and how I lost my health (which is almost everything) because of the stinking smoke [*Kinrauch*] and the horrible vapors."[69] Chrysophilus is sure that his friend

Theophrastus has already heard about his misfortunes, "how disgracefully I have been deceived and set upon [*angesetzt*] by dissolute lads, who claim to be alchemists and the disciples of philosophers." Up to this point, there is nothing particularly remarkable about this dialogue, which resembles earlier critiques of alchemy (Erasmus's dialogue about Balbinus, for instance). Then the dialogue takes a different turn, however, as the two go on to discuss the differences between "natural alchemy," which relies on the "secrets of nature" for its success, and the type that depends on "mere fraud [*lauter Betrigerey*]." "The notion that those same vagabonds and frauds [*Landstreicher* and *Landbescheisser*] (whether they are learned or not) could be at all helpful to you," Theophrastus admonishes his friend, "is as believable as the notion that one could make ivory white with something black."[70] Theophrastus concludes the dialogue (and the entire book) by teaching Chrysophilus how to recognize "not only false sophistical books, but also those who deceive the entire world [*Landtbetrieger*]" by their verbosity, obfuscating language, clichéd phrasing, and frequent requests for more and more equipment.[71] The repeated mention of deceptive alchemists in this text promoting alchemy illustrates how central the issue had become for alchemical authors at the end of the sixteenth century.[72]

Books like the *Alchimyspiegel* aimed to convince readers that they must learn to differentiate real alchemy from alchemical *Betrug*. A few authors, however, also used the issue of fraud more constructively to articulate a new dimension of the alchemist's persona: the alchemist as a discriminating expert. As Eric Ash has argued recently, a whole range of "expert mediators" came to play an important role in sixteenth-century state building, both in Elizabethan England and in Europe as a whole. Alchemists too hoped to take advantage of this trend, positioning themselves alongside those who could coordinate and access all of the other practitioners necessary to statecraft, from shipbuilders to mapmakers.[73] The physician and alchemist Leonhard Thurneisser, for instance, utilized this strategy in the preface to his treatise on Paracelsian medical alchemy, *Magna alchymia* (1583). Thurneisser took an anecdotal approach, linking a more abstract discussion of the qualities of the ideal alchemist to stories of alchemical deceit that had already been circulating for decades. His aim, he claimed, was to warn and edify potential patrons about the possible pitfalls of investing in alchemy. Because those of high station "are unaccustomed to the damaging fumes (which issue from metals and minerals, as well was as from the waters and oils and other things prepared from them)," Thurneisser noted, "and also the work of the fire, the soot and the distillation [*Lutirens*], the watching and the dirty work [*Wachens und Sudlens*]," they prefer to hire servants or *Laboranten* to perform these tasks for them. Thurneisser did not hide his disapproval of farming out the labor of this most noble art to others, but he clearly also recognized that this very same practice provided him with noble support for his own work throughout his life, so he stopped short of condemning the practice altogether. Thurneisser instead turned to a discussion of the qualities of "the highest virtues that such a *Laborant* should have."[74] By using the term *Laborant*, or "laboratory worker," he indicated his particular focus on entrepreneurial alchemists who would work for wages. Above all, the alchemical employee should be of good moral character: true to God ("and not only in appearance," but deep in his heart), not a drunkard, gambler, or gossip. He should be honorable, discreet, peaceable (*inmütig*), and brave, but not proud, insolent (*übermütig*), or the type to dress above his social rank just so that powerful people will revere him. He should be watchful, cautious, diligent, and earnest, not "frivolous [*leichtfertig*] in manner, appearance or gestures." Perhaps alluding to Petrarch's *De remediis*, Thurneisser insisted that the *Laborant* should remain stable and constant when faced with both good fortune and hardship alike, rather than becoming too exuberant in times of luck, or "beaten down, weak, effeminate, or fearful" in times of adversity. Thurneisser also pointed out the obvious: any alchemical assistant one hired should be knowledgeable about metals, minerals, and the fire, smelting, assaying, mining, and so on. Finally, this ideal alchemist should be willing to endure any kind of smoke, vapors, heat, cold, hunger, thirst, or bad weather—in other words, the physical trials of alchemical work that wealthy employers preferred to avoid.[75]

These clusters of qualities would have been mostly familiar to connoisseurs of alchemical literature, for Thurneisser's was a fairly typical discussion of the alchemist's virtues. In the next section of his preface, however, Thurneisser elaborated on the theme of *Betrug*, delineating four different types of deceitful alchemists in general "against whom princes who love the arts and other honorable people should guard themselves most carefully."[76] The first type of "wandering, false, *betrüglicher Alchymist*," according to Thurneisser, roams the Holy Roman Empire decked out in "large chains, as well as gold-plated weapons and daggers made of white copper or some kind of mixture, and golden rings set with" all kinds of imitation gems.[77] These glittering vagabonds stay with wealthy innkeepers who have contact with princes, gradually forging connections with prominent people. At first, they are quiet and humble, and when asked about their business, "they answer briefly, that they were poor journeymen, but that God bestowed good fortune upon them, so that they may be satisfied; they desire nothing more than to live out their life in peace." Few can resist this bait, Thurneisser continues, and soon many grovel to learn the alchemists' secrets, lavishing them with gifts, and eventually investing money. Meanwhile, the alchemists stall, collecting more money to send out for more materials, a powder or an ore, a stone or a root; then, when the time is right, they disappear. *Betrüger* of the second type, in contrast, arrive dressed in rags, wearing a garment so worn that "not a single louse crawls on it." They explain that they were robbed, or in a shipwreck, and that all of their books and clothes were destroyed. Fortunately, they explain, God delivered them into safety and blessed them with a gift with which they can provide for themselves and make others rich too. Once again, people rush to bestow gifts and money on these false alchemists in order to learn their secrets, and "so they receive clothes, and fill their bags with money, and they walk off." The third type offers his art for nothing but the necessary food, lodging, and materials while he works; in exchange, he promises to grow gold or silver "as large as a hundred-year-old oak tree" out of a initial mixture of gold, silver, and several other ingredients. He explains that no matter how much one starts with, this process requires the same amount of time and work; inevitably, the patron, eager to produce as much gold as possible at one time, insists on doubling or even tripling the initial ingredients in order to make more precious metal. The alchemist combines the ingredients (secretly holding back the silver and gold), then claims that he needs to go and purchase one more ingredient in order to succeed. Leaving the mixture on the fire, he takes the gold and silver with him and (once again) disappears. Finally, the members of the fourth "sect [*Sect*] of these false *Laboranten*" let it be known that they know many secret things and what they need to carry it out. Once enthusiasts put up some money to acquire the art or carry it out, "they do as the others" and leave.[78] For Thurneisser, the initial presentation of all four of these alchemists was the heart of their deception. False words and false clothes, whether excessively lavish or overly wretched, were meant to create the trust that would prompt the wealthy to willingly turn over their money.

Thurneisser's description of the four types of *Betrüger* took a slightly different tone from his initial, more abstract musings on the qualities of the ideal alchemist. In his discussion of fraud, Thurneisser's text was more detached and descriptive, suggesting that he was merely reporting the state of affairs as they were. "And everyone can look into all of this for himself," Thurneisser recounted; "In thirty years (while I was in Tarrentz and Berlin), I have [seen] a whole host of deceivers [*ein gantz Feinlein*] who were fed and clothed by five hundred nobles and people of high station, even though they [the deceivers] were lying and deceitful knaves."[79] And yet it would be far too simple to conclude from this particular rhetoric of witnessing that Thurneisser was simply reporting what he had seen. After all, more obviously fictional accounts of alchemical fraud, such as Erasmus's colloquies, were also narrated in the first person, a hint that this was not just a report but a rhetorical strategy. Thurneisser may well have encountered *Betrüger* in his day, but he certainly also encountered their literary counterparts in the pages of Brant or Agrippa; more important, he knew that so too had his readers. By seeming to authenticate literary reports of alchemical *Betrug* by attributing them to his own personal experience, Thurneisser bolstered his own authority on the matter of alchemy. Alchemical *Betrug* was no mere literary fiction, Thurneisser implied, but a very real threat that any well-intentioned patron might face. "Anyone who wants to practice this art, whether he be of high or low social rank . . . should think this over carefully," Thurneisser concluded his preface, "and learn to attend to [*observiren*] and recognize such fellows . . . now and before he gets involved with them."[80] Alchemists like Thurneisser, of course, provided the solution at the same time that they authenticated the problem. They offered themselves—or at least their books—as precisely the experts that anxious patrons needed in order to avoid deceitful alchemists. As we shall see, some patrons took such experts up on their offer.

The way in which alchemical authors appropriated stories of fraud to bolster their own authority was even more apparent in the 1597 "Treuhertzige Warnungs-Vermahnung eines Getreuen Liebhabers der Warheit an alle wahre Liebhaber der Naturgemässen Alchymiae *Transmutatoriae*; daß wegen der Bübischen Handgriffe der betriegerishcen Arg-Chymisten gute Auffacht vonnöthen" (Heartfelt warning and admonition from a faithful devotee of the truth to all true enthusiasts of the natural transmutational alchemy, which necessitates close attention because of the roguish tricks of the wicked *chymists*). Khunrath was clearly the author of this short tract on fraud, which was published with his much longer treatise on alchemical philosophy, *Von hylealischen . . . Chaos* (On primordial chaos, 1597). Yet Khunrath distanced himself somewhat from the "Warnungs-Vermahnung."[81] Rather than claim authorship for himself, he ascribed it to one "Thrasybulus Ricenus," a thinly disguised Greek and Latin version of his own name. Khunrath removed himself from the "Warnungs-Vermahnung" further by noting in the *Chaos*, "The deliberately fraudulent goldmaker's knavery that is printed at the end of this work is neither prejudicial nor unfavorable to me; it was penned as a separate tract and does not belong to my confession [i.e., the *Chaos*], as its separate publication clearly shows."[82] Khunrath's choice to hide his identity is somewhat puzzling, especially given the fact that he addressed the issue of fraud so openly in other works in his own name.[83]

If Khunrath hid behind Thrasybulus Ricenus in the byline for the "Warnungs-Vermahnung," he certainly did not conceal his views on alchemical *Betrug*. "My dear friend," he began, "follow my heartfelt and extremely useful advice: if a gold beetle comes flying over to you, saying that he knows how to make gold and silver and that he wants to teach you, do not believe him easily or hastily, because this is not as mean an art as many let themselves dream it to be." The short tract went on to detail over forty-six different sleights of hand that *Betrüger* purportedly carried out, including the tricks with the hollow coals and stirring rods, double-bottom vessels, switching glasses, and so on, as well as how to safeguard against alchemical cons.[84] The author urged his readers to take alchemically produced gold to a trusted assayer for testing, for instance, or to purchase the materials for the alchemist's work "in various cities miles apart from each other and from different merchants, in the east, west, north, and south."[85] Finally, lest the author himself be confused with those *Betrüger* who sell their alchemical secrets for money, he assured his readers that he wished only one thing in exchange for his advice: that good people shun "the Evil, *Algemistische,* and intentionally fraudulent guild of gold beetles" (*Algemistische,* again, being a pun meaning "all is dung"), taking them "for nothing other than evil-smelling and foul apes. Everyone who sees one should spit at him and say: Fie, you fraudulent goldmaker, AMEN."[86]

Like Thurneisser, Khunrath described alchemical *Betrug* as firsthand experience. He presented his "deceitful gold beetles" not as metaphors or vehicles for moralizing, but as observations of common events in his own day. "Daily experience proves this," Khunrath assured his readers.[87] And yet readers familiar with the literature on alchemy would have recognized many of these tricks as commonplaces from Petrarch, Chaucer, Erasmus, or other earlier texts. By adopting a rhetorical strategy of presenting old literary tropes as common experience, the author walked a fine line; perhaps this too explains why he chose to write as Thrasybulus, rather than as Khunrath. As he acknowledged, one could easily wonder how he had come to know so much about these tricks. "Now should I, or anyone else . . . be an Evil-Chymical *Betrieger,*" he asked, "just because we faithfully bring to light and describe this fraudulent knavery as a useful warning to honorable people?" If describing *Betrug* makes one a *Betrüger,* he argued, then all of the previous alchemical authors who have addressed the issue, from Bernhard of Treviso and George Ripley to Dionysius

Zacharius, must also be *Betrüger.* By the same logic, any theologian who preaches against adultery must be an adulterer; any physician who teaches about disease must be himself diseased; any jurist who lectures on the proper punishments for theft must be a thief. In fact, Khunrath ventured, should not God himself then be a murderer, adulterer, magician, and sinner, since the Bible discusses such issues at length?[88] With this comparison, he carefully removed the argument that he was a *Betrüger* just because he revealed their secrets. At the same time, by comparing himself to other experts (theologians, jurists, and physicians), Khunrath constructed a position of authority from which he could evaluate and resolve the problem of alchemical fraud.[89]

This strategy resembled that of the Italian "professors of secrets" whom William Eamon has described from this same period. These hybrid figures, too, took advantage of both a burgeoning print market and their liminal position between official academic culture and the piazza or workshop to publish the artisanal "secrets of nature" that would appeal to elite audiences: the medical secrets of popular pharmacology, techniques for making soaps, perfumes and scented lotions, as well as inks, dyes, and artificial gemstones. These authors positioned themselves as experts who could divulge both the artisanal secrets hidden in workshops and the secrets concealed by nature.[90] Similarly, alchemists like Thurneisser and Khunrath cast themselves as people with unique insight into the fraudulent deceptions of the *Betrüger,* divulging their tricks and warning patrons that they were virtually under siege by the legions of alchemical swindlers swarming the Holy Roman Empire. At the same time, they took control of the alchemical persona again, acknowledging and condemning their disgraceful cousins, the *Betrüger,* and putting forward the alchemist as expert in their stead.

By the early seventeenth century, the alchemist's persona had absorbed a number of elements that were not present in the late Middle Ages. Although the prophetic dimension of alchemy largely (but not entirely) receded in the sixteenth century, the alchemical persona still incorporated the elements of both the scholar and the artisan that had long been a part of alchemical tradition. Alchemists, in fact, came to epitomize for many the union of hands and minds that drove new ways of thinking about nature in the seventeenth century. At the same time, however, alchemists now had to contend with their doppelgänger, the false alchemist. This figure, sketched largely by literary observers and artists who were not involved directly with alchemy, had an important, if paradoxical, effect on the development of alchemy in the sixteenth century. On the one hand, the witty prose descriptions and images of the *Betrüger* damaged alchemy's already shaky reputation by associating it with foolishness, dishonest trade, and even crime. On the other hand, these images developed this persona in ways that finally made the alchemist seem to be a part of the early modern social world. Europeans could grasp the alchemist as *Betrüger* as a social type and situate this new character among the fools, merchants, and rogues that had come to be stock characters in the imaginary landscape of early modern Europe. By making the *Betrüger* a familiar character by the end of the sixteenth century, paradoxically, these authors and artists paved the way for the alchemist to succeed as a real persona in early modern Europe. Alchemists now had a recognizable foil against which to define themselves and differentiate the alchemist's persona. The *Betrüger,* in other words, made the persona of the real alchemist as discriminating expert necessary, relevant, and comprehensible to central European elites who did not want to fall prey to the transparent schemes of the false alchemists.

But what relationship did such literary and visual images of the alchemist have to the real-life alchemists who populated the cities and courts of the Holy Roman Empire? Were there really *Betrüger* who provided the models for Petrarch and his followers? Historians of alchemy have not devoted enough critical attention to this complicated question, tending either to simply ignore such images, perhaps regarding them as merely the irrelevant uninformed observations of outsiders, or viewing them (especially the visual sources) as reflective of a real world of sophists (or puffers), alchemical charlatans, and true adepts.[91] I have argued that these texts and images are in fact crucial to any understanding of the history of alchemy in this period, not because they can be used transparently as sources that *reflect* what alchemists were really like, but, quite to the contrary, because they helped to *create* a meaningful persona that was a central part of early modern alchemy. The representations of alchemy in Brant, Brueghel, and their peers asserted surprisingly complicated and nuanced arguments, however critical, about the social and cultural significance of alchemy; in doing so, they made the alchemist seem realistic and meaningful as a social type. In this sense, these texts and images were no less crucial to the development of alchemy in this period than the writings of natural philosophers or the activities of alchemists.[92]

Nevertheless, personae are constructed not only in texts, but also through the habits and self-presentation of individuals. As Daston and

Sibum remind us, after all, the persona is "a cultural identity that simultaneously shapes the individual in body and mind and creates a collective with a shared and recognizable physiognomy."[93] It remains, then, for us to understand how the elements of the alchemist's persona I have outlined here shaped the lives of both the alchemists who had to inhabit it and the expectations of the patrons who hired them.

CHAPTER THREE

Entrepreneurial Alchemy

With the halting emergence of various personae in the sixteenth century, alchemists became recognizable, if contested, figures in the social and cultural landscape of the Holy Roman Empire. Practitioners' lives were not shaped solely by persona or knowledge and skill, however, but also by a third factor: the motivations and interests of the patrons who supported alchemists and their work. Although there are hints that urban patricians and members of the clergy were numerous and important as patrons of alchemy in the early modern period, princes and merchant capitalists were by far the most visible supporters of alchemical work. In large part, these patrons fueled the market for alchemy in the Holy Roman Empire by creating an enormous demand for it; the large number of princes (and thus potential patrons) in the Holy Roman Empire, in fact, goes a long way toward explaining why alchemical practitioners flourished there, as these patrons built laboratories and hired alchemists to carry out all kinds of work in the sixteenth and seventeenth centuries. The appearance of alchemy at early modern courts, in the newsletters of the Fugger banking family, and in the correspondence of elites raises a set of important questions about its widespread patronage. Why was alchemy of such interest to the political and financial elites of the Holy Roman Empire? Was it simply that only the wealthiest Europeans could afford to pursue this expensive art, or was there something about alchemy that particularly resonated with the political, economic, or even religious concerns of Europe's elites? Moreover, of what consequence was it that the demand for alchemy came largely from the most powerful ranks of the empire? How did this shape the practice of alchemists like Philipp Sömmering?

In some ways, these are not new questions. Historians have long noted the prominence of alchemy at princely courts and have offered a variety of explanations for alchemy's particular appeal to central European elites in the late sixteenth and early seventeenth centuries. A century ago scholars condemned sovereigns' interest in natural magic or alchemy as a kind of escapism, a flight from truly important affairs of state while the Holy Roman Empire headed toward the Thirty Years' War. More recently historians have defended these princes' pursuit of "occult philosophy," or *philosophia hermetica,* arguing that in the sixteenth century such philosophical interests were in fact a meaningful intellectual response to religious and political crisis. In the face of religious conflict, war with the Ottoman Empire, the rise of nation-states, and the first signs that a total philosophy of nature was fracturing into separate disciplines, European rulers such as Holy Roman Emperor Rudolf II or Landgrave Moritz of Hessen-Kassel turned to natural magic and hermetic philosophy, where they could locate a comforting unity to the world more fundamental than widening surface fissures. The philosophical universalism and irenicism characteristic of these philosophies posited a single, universal natural order that linked the entire cosmos without regard for nation or confession, offering a worldview strikingly different from the divided European reality. Bruce Moran has described the appeal of this view most poetically. "The occult vision of unity and universality," he argued, "offered an intellectual balsam for religious and political confusion. As such, it became a surrogate reality, and it is in this sense that its patronage, as much at Hessen-Kassel as at other German courts, became finally a patronage of despair." This "intellectual balsam" was not merely symbolic. From the perspective of natural magic, which linked everything in the cosmos through infinite and invisible correspondences, knowledge of how to control nature could be just as powerful as seemingly more concrete forms of power (such as political control or warfare). As part of a broader intellectual effort to understand and control nature, therefore, natural magic and hermetic philosophy offered central European princes a solution to pressing political and religious problems. After all, Emperor Rudolf II, whose court has come to be emblematic of this cosmology, fought the Ottoman Empire as much with astrological forecasts and paintings of European victory as he did with military campaigns.[1]

Historians have tended to subsume alchemy within this more general "occult philosophy," focusing on the power that alchemical ideas could offer early modern elites. Because alchemy was both a philosophy of nature and a practical technology, however, it offered more direct and concrete opportunities for control as well, especially in the economic arena.[2] As this chapter argues, alchemists' practical expertise in the extraction and refinement of precious metals made them valuable advisers in the waning years of the sixteenth century to an extent that has not yet been fully appreciated. Although princes and others invested in the mining industry recognized a difference between mining technologies such as smelting and assaying and alchemical transmutation, they saw them as related technologies that could improve mining industries and make them more profitable. Alchemy was attractive to political and financial elites not just as an idea, in short, but also as a technology. Alchemists' increasing involvement in mining had profound consequences, however, as princely expectations about profits and yields shaped their perceptions of alchemists' abilities.

"This Metallic Blessing": From Boom to Bust

In 1593, Conrad von Grumbach offered the Würzburg prince-bishop Julius Echter von Mespelbrunn a secret process for making gold. The bishop agreed to sign a contract with the alchemist, offering Grumbach the villages of Rimpar and Bergtheim (worth approximately two hundred thousand gulden) in exchange for the process. The bishop's stated reason for buying Grumbach's recipe was quite simple: the alchemical production of wealth would serve the Catholic Church. As he put it, "because the Almighty does not give just anyone this grace and gift, which is not trifling but singular, but only [gives it to] one who dedicates it to the heavenly Almighty, we should and desire to use it not only for the praise and honor of his name, but also for the propagation of his godly and beatific word, preservation of the Christian [i.e., Catholic] church, good peace, and our land, people, and poor subjects."[3] By dedicating his promised alchemical riches to the Catholic cause, von Mespelbrunn simultaneously justified his desire for gold and his own position of authority. Since God bestowed his gift only on the worthy, the success of his transmutations would not only enrich the bishopric's coffers, but also indicate God's favor of both Mespelbrunn and the Catholic Church.[4]

Like the prince-bishop of Würzburg, some early modern rulers expressed the hope that alchemy could fund various kinds of political projects. Without question, the territorial states of the empire faced growing costs in the sixteenth century. Many rulers simply needed an additional source of income to make up for budgetary shortfalls, and alchemy's promise of transmutation could easily seem to offer a solution. This financial crisis had deep roots; since at least the fifteenth century, the princes, nobility, and clergy of the Holy Roman Empire had had difficulties generating enough revenue solely from their traditional sources of income in an increasingly commercial economy.[5] These challenges only increased in the sixteenth century. Debts accrued as a result of war, Renaissance building projects, and new expenses associated with the implementation of religious reforms and state centralization. Emperor Charles V's debts upon his abdication in 1556, for example, were about as large as his income from Spanish America during his entire reign.[6] Likewise, when Elector August of Saxony took the reigns of power in 1553, he inherited a debt of 1,667,078 gulden; this debt had piled up in part, no doubt, because August's brother and electoral predecessor, Moritz, had been paying sixty-four thousand gulden a month to employ soldiers to fight the ongoing wars of religion in the empire—an expense that August saw fit to continue despite the accumulated debt.[7] Alchemy was not the only option for princes who could not balance the books—many simply took out loans from prominent banking families in Augsburg and Nuremburg—but it certainly would have been an appealing alternative.

Reducing princes' economic interest in alchemy solely to a desire to make gold quickly, however, would miss the much more interesting and sophisticated relationship between alchemy, mining, and finance in early modern central Europe.[8] Early modern princes' involvement with alchemy must also be understood as part of their long-standing engagement with the mining industry as a source of income for the state. Although princes had long depended on the mineral resources in their lands as a source of income, their relationship to the mining industry as both sovereigns and investors intensified in the mid-fifteenth century. After a long period of decline in the fourteenth and early fifteenth centuries, increasing demand for metals (for both guns and currency) focused attention in the mid-fifteenth century on the possibility of reopening stagnant mines and locating new ones.[9] Technological innovation seemed to offer the key, as the largest problem was the fact that miners had so far been unable to mine ore at depths below the water line. Responding to this challenge, engineers developed new water pumps, ventilators, and systems of haulage that enabled miners to build deeper mine shafts and reach hitherto untapped sources of ore.[10] Innovations on the smelting side of the mining industry improved productivity as well. A metallurgical innovation known as the *Saigerprozess*, for example, made it possible to separate silver from copper ore through the use of lead. These technological developments created new possibilities for the mining industry in the second half of the fifteenth century.[11]

These new technologies were also extremely expensive, however, and soon the need for capital to take advantage of them spurred innovative financial arrangements that drew in merchants and princes as investors.

For the small-scale collectives that had traditionally worked mines and processed ore, sharing in the profits together, the new capital-intensive machines were simply beyond reach. A piston pump, for example, cost five hundred to seven hundred gulden, while a *Kehrrad,* a hydraulic wheel for removing water from mines, could cost three thousand gulden. The *Saigerprozess,* meanwhile, required two-thirds more capital and labor than any other contemporary process, largely because of the higher fuel costs and necessary equipment such as the embossing hammers, or *Tiefhämmer.*[12] In order to attract the capital to fund new mining operations, therefore, mining companies offered shares, or *Kuxen,* to outside investors, essentially exchanging a share of the future production for an advance of capital to buy new equipment.[13] Eventually, some investors got involved in smelting and forging works as well as mining, finding opportunities in both the extraction and the refinement of ore. The Fuggers, for example, built three huge smelting houses in Upper Hungary, Thuringia, and Carinthia in order to process the ore they mined.[14]

These new financial arrangements altered the traditional relationship between princes and the mining industries in their territories. As rulers, they had long held the rights, collectively known as the *Bergregal,* to mine the minerals in their lands themselves or to grant those rights to mining companies in exchange for taxes.[15] At the end of the fifteenth century, however, princes also became investors in those mining companies. Duke Georg of Saxony (1471–1539), for instance, owned seven hundred *Kuxen* distributed across the forty mining works in the Annaberg district of Saxony.[16] He also became a shareholder in a tin-producing company in 1491.[17] Princes' new role as investors gave them a new motivation to ensure that companies succeeded, and they drew on the resources of the state to do so. Rulers began to promote mining, for example, by bringing in expert advisers to improve local enterprises. Thus, in 1524 the Wolfenbüttler Duke Heinrich the Younger (1489–1568) brought the former Joachimsthaler *Bergmeister* [mine overseer] Wolf Sturz to run the Harz mines in Heinrich's territory. Sturz brought miners from Joachimsthal with him, thus increasing the store of expertise in Heinrich's territories.[18] Local rulers sometimes invested in technology as well. The count of the copper-rich region of Mansfeld, for instance, brought Nuremberg-built furnaces for refining copper to Schleusingen in 1461 and to Hohenkirken in 1462.[19] Rulers could also grant privileges, of course, and they took advantage of this right to grant monopolies to companies in which they themselves held shares. In 1533, for example, Kurfürst Ludwig V von der Pfalz created a company, in which he was a major shareholder, and then guaranteed that it would be profitable by granting it the privilege to purchase all of the tin produced at Amberg. The Saxon Duke Georg set up a similar arrangement with a tin-producing company in the 1490s.[20] As these kinds of arrangements demonstrate, princes played overlapping roles in the mining industries in their lands. They held regalian rights as well as shares in mining companies. Moreover, by creating new administrative positions for mining experts and granting privileges to bring in merchant capital, princes drew on the power of the state to promote and develop the mining industry.

The new technologies, infusion of merchant capital, and princely promotion of the mining industry proved to be a powerful combination; the output of silver, copper, and other minerals from central European mines increased phenomenally from the 1460s to the middle of the sixteenth century.[21] Yields varied from region to region, of course, but generally speaking, mining output of silver and copper increased five- to sixfold by the 1520s.[22] As investors, princes and bankers such as the Fuggers benefited substantially from the mining boom. In Saxony, for instance, income from silver mines accounted for a quarter of the state revenue in 1470, and two-thirds in 1530.[23] The profitability of the Fugger mines in Hungary and the Tyrol at the beginning of the sixteenth century was astounding, funding the Fugger bank and notoriously enabling them in turn to fund the bribes ensuring Charles V's election as Holy Roman Emperor in 1519.[24] All told, European mines were producing about ninety metric tons of silver annually by the mid-1520s.[25]

By the mid-sixteenth century, however, European mines stagnated again and the burst of productivity began to level off. In central Europe, silver output dropped by almost half in the second half of the sixteenth century, from an average annual production of 19,850 metric tons in 1545 to only 10,400 metric tons in 1620. Production varied from region to region, of course. In Saxony, the drop was particularly dramatic, falling from 13,150 metric tons in 1545 to 5,100 metric tons in 1620. In the Upper Harz near Wolfenbüttel, on the other hand, yields actually went up slightly during this period, from 3,400 metric tons in 1545 to 4,500 metric tons in 1600, and back down to 3,550 in 1620.[26] The Fuggers, heavily invested in mining industries in Austria and Hungary, faced a similar crisis. By 1550, in Schwaz in the Tyrol, silver production had declined by one-half since 1523;[27] by 1570, the productivity of the Carinthian mines had decreased significantly, forcing the Fuggers eventually to sell one of their smelting works, the Fuggerau.[28]

Just as the central European mines stagnated, gold and silver from Africa and the Americas began to flow into Europe. The Portuguese imported about 36–40 metric tons of African gold (worth about 520 tons of silver) in the early part of the sixteenth century.[29] Europeans began to exploit sources of American silver in Mexico in the 1540s and 1550s and in Potosí, Peru, in 1543. The Spanish introduced the mercury amalgamation process in Mexico after 1556 and in Peru after 1572, along with draft labor systems that facilitated the exploitation of native sources of labor; as a result the extraction of silver from ore became even more productive, increasing further silver exports to Spain.[30] American silver exports to Europe increased more than thirtyfold in sixty years, in fact, surging from 86 metric tons in the 1530s to 1,118 in the 1570s and 2,707 in the 1590s.[31] These global shifts in supplies of precious metals interacted with dramatic changes in the European economy in the second half of the sixteenth century. The massive influx of American silver caused inflation in the European market, making German silver less valuable and further diminishing the profitability of already failing German mines. At the same time, the European population recovery in the mid-fifteenth century after the demographic devastation of the Black Death, together with increased overseas trade and improved internal European networks, contributed to expanding markets. By the mid-sixteenth century, in other words, the European economy was booming but the increased economic activity required ever more silver and gold, much of which now came from the Americas rather than central Europe. After about 1590, however, American silver imports began to decline as well, causing a slowdown in the European economy and a gap between the level of economic activity and the amount of specie available to facilitate it.[32] In short, the booming economy required more precious metal at the end of the sixteenth century, precisely when central European (and eventually American) supplies were diminishing. As the economic historian F. C. Spooner explained, "at a time when the economy continued to expand, even at a slower rate, and required increasing quantities of money, the supplies began to slacken and then decline. There was an overriding need to supplement the existing facilities, to adopt a new attitude."[33]

Princely Entrepreneurs

These broad trends posed a serious challenge for the princes and merchant bankers of central Europe because they were so directly invested in the interconnected world of mining and finance. Just as they had responded a century earlier by encouraging technological and financial innovation, princes and merchant bankers took action once again at the end of the sixteenth century. One response at the imperial level was to try to keep gold and silver from leaving the Holy Roman Empire; the Imperial Assembly passed decrees in 1524, 1555, and 1570 restricting the export of gold and silver coinage from the empire.[34] Another response was to mint coinage of baser metals—copper-silver alloys or pure copper, rather than silver; these coins began to circulate at the end of the sixteenth century.[35] A "veritable explosion in credit" in this period also helped to solve the problem of limited specie. Banking houses, most prominently the Fuggers in Augsburg, were able to use the bill of exchange to expand economic activity beyond the actual amount of available gold and silver.[36] Finally, at the territorial level, princes continued to encourage economic innovation and development.[37] The identification, exploitation, and production of new natural resources became central to the economic policies of late sixteenth-century princes like Duke Julius of Braunschweig-Wolfenbüttel, Elector August of Saxony, and Duke Friedrich of Württemberg.

Duke Julius of Braunschweig-Wolfenbüttel took an interest in his territorial economy in the late sixteenth century as both a sovereign ruler and as a *fürstliche Unternehmer*, or princely entrepreneur, who invested in the mining industry in order to make a profit. Like his father, Duke Heinrich the Younger, Julius personally invested in mining technology and production in the rich copper and silver mines in Goslar and the Upper Harz Mountains. He owned some entire works outright, held shares in others, and also collected regalian dues and taxes on the mining industry, which was by far the largest source of ducal income in this period.[38] His interest in making the mines more profitable, therefore, stemmed from both his roles as individual investor and territorial sovereign.[39] At the heart of Julius's mining policy was a campaign to catalog and develop the natural resources in his territory. He hired a series of experts with experience in other mining regions as consultants. In 1572, for example, Julius ordered his administrators to assist the surveyor Gottfried Mascopius in completing a survey (*Landtafel*) of the territory with an eye to its potential economic development. That same year, Julius's adviser Erasmus Ebener, who came from Nuremberg, completed a report on "all kinds of mountains, metals, and whatever else is useful, which are found in [the mining regions of] the Harz and especially Rammelsberg."[40] Ebener's report demonstrated a clear concern with the profitability of the mines and related smelting processes. Regarding the gold deposits in Rammelsberg, he noted: "All of the silver in Ramesperg [*sic*] contains gold. One mark contains on average five *Mariengroßen* of gold; should the gold be extracted, it will not create much profit. There is an art that recovers more, [but] one doesn't know how much [more], since it is not commonly assayed and smelted."[41] Fourteen years later, in 1586, Julius hired the *Bergmeister* Hans Fischer from Heidelberg to search the ducal territory for natural resources.[42] Such reports were part of an effort not only to identify natural resources in Braunschweig-Wolfenbüttel, but also to evaluate their potential economic value.

Hans-Joachim Kraschewski has argued that Julius's efforts to improve the local mining industry were connected to a much broader political and economic reform program. Julius approached the mining industry in part from an administrative perspective, reorganizing both the *Landesherrliche Verwaltung der Berg- und Hüttenwerke* (Territorial Administration for Mining and Smelting Works) and *Verwaltung der Forsten im Harz* (Administration for Forestry in the Harz), which provided fuel for the smelting works that processed Harz copper; these were part of a more general administrative overhaul designed to streamline and centralize the state bureaucracy.[43] Duke Julius also cultivated the local mining industry by extending advances and credit to mining and smelting companies operating in his territories. As Kraschewski has argued, the duke sought to make the mining industries in his territories more profitable so that they would bring more capital into the ducal treasury through taxes. Julius planned to reinvest this capital in local industries, providing raw materials and credit to artisans who were struggling to keep up with the changing economy. A more active trade with neighboring Hessen, Julius hoped, would also stimulate these local industries. As he explained in a letter to Landgraf Wilhelm IV of Hessen in 1576, the two rulers should work to intensify this kind of bilateral trade, "and therefore to create more *Comertia* and trade [*Handtierung*] between Your Dearest's lands and mine and to mutually advance our subjects."[44] The ultimate aim of Julius's economic policy was to use the profits from the mining industries not only as a direct source of income, therefore, but also as a way to generate capital and then use it to subsidize local industries and protect them from the general inflation of the sixteenth century. At the same time, he also centralized state control over the wide range of economic activity in his territory.[45]

Duke Julius was not alone among sixteenth-century princes in this type of economic innovation. Elector August of Saxony has been described as "the most important early capitalist entrepreneur" as well as "one of the most successful economic policy makers [*Wirtschaftspolitiker*] on the German princely throne," as evidenced by the fact that he transformed a debt of over one and a half million gulden at the beginning of his reign into a profit of over two million gulden ten years later.[46] He did this in large part by investing in Saxon agriculture, a focus that grew out of August and his wife Anna's very personal interest in nature. Their pursuit of natural knowledge and medicine was reflected in their extensive collections. The electoral library contained 2,354 books of August's and 450 books of Anna's by 1580, many of which dealt with practical matters like horticulture, horsemanship, medicine and alchemy, mining, and mathematics; these books seem to have been largely for the personal use of the elector and electress. The *Kunstkammer,* meanwhile, contained not only more books, but also a number of other objects to do with the measurement of nature: clocks, globes, astrolabes, compasses, and divining rods. August actually made several maps of his territories himself, evidence that the surveying tools in the *Kunstkammer* were not merely for display.[47] August and Anna also had a particular interest in gardening and horticulture. Their correspondence shows that the electoral couple exchanged information, seeds, and plants with other courts. In 1560, for example, August wrote to Duke Christoph of Württemberg to ask him not only to send seeds of nonnative herbs, fruit, flowers, and other plants, but also to allow August's gardener Nickel to enter into an exchange with the Stuttgart gardener Coßmann von Metz. The Württemberg duke obliged, and a couple of months later Coßmann von Metz sent Nickel 170 scions, or grafting shoots (*Samen-Pfropfreiser*).[48] Anna, who was widely known for her medical expertise, maintained her own botanical garden, in which she experimented with medicinal herbs. August shared his wife's interest in gardens, albeit from a slightly different perspective. August took a particular interest in fruit cultivation; he learned how to graft trees, exchanged Saxon grain for new varieties of fruit trees, and eventually recorded his knowledge in a book entitled *Künstliches Obst- und Gartenbüchlein,* which was published in 1571 and reprinted multiple times.[49] His own planting tools were preserved in the electoral *Kunstkammer,* along with a note in the inventory indicating that he actually used them for planting seeds. As Alisha Rankin has argued, the activities of this electoral couple demonstrate an extensive, hands-on, experimental approach to the manipulation of nature that surely places *both* August and Anna among those whom Bruce Moran has described as the prince-practitioners of the early modern period.[50]

The electoral interest in nature in Saxony was not merely a personal hobby, but also central to the political and economic power of the court. Like Emperor Rudolf II, Moritz of Hessen-Kassel, and the other princes of the Holy Roman Empire, the electors of Saxony drew heavily on imagery from nature (and mining in particular) to legitimate their reign through court festival and cultural productions. As Helen Watanabe-O'Kelly has pointed out, the fact that "the theme of mining is presented again and again in court festivals is a phenomenon peculiar to Dresden." In a carnival tournament in 1574, for example, Elector August himself appeared as Mercury, "God of all Metals." He offered protection to the miners, who also appeared with realistic tools of their trade—miners' lamps, caps, divining rods, ore, and a miner's axe—and were joined by a smelter and a coin maker with their tools. These kinds of tournaments, which continued to utilize mining themes into the seventeenth century, emphasized how absolutely central mining was to both the political and economic fortunes of Saxony.[51] As Jutta Bäumel has noted, "it is fair to say that with his activities as a craftsman and his research Elector August not only pursued a personal passion but an entire economic agenda."[52] Among other things, August introduced new types of grain and cattle to Saxony, promoted horse breeding and founded forty sheep farms, all of which were designed to relieve the elector's debt. He also purchased twenty-seven new farms in the first twenty-five years of his reign, which brought in new income and served as models for how estates should be cultivated. Model farms in Ostra and Stolpen, for example, provided seedlings for sale or exchange and demonstrated new grafting techniques that could be used to propagate particularly pest-resistant species. August's interest in new types of crops led to, among other things, the introduction of the potato to Saxony in 1591. Anna, meanwhile, also experimented with dairy products on these farms, which she "made commercially viable in Saxony for export to her native Denmark." All of these projects proved to be wise investments for the electoral couple, bringing 250,000 gulden annually into the treasury by the 1560s, and by the end of his reign, approximately 400,000 gulden yearly.[53]

Like agriculture, mining was also something that interested Elector August both personally and for its role in political economy, and he devoted particular attention to developing mining as a central industry in his territories. In part, like Julius of Branschweig-Wolfenbüttel, August sought to reorganize and gain control over the industry through a series of new mining regulations (*Bergordnungen*) in 1554, 1571, and 1575. These new regulations not only centralized the mining industry and consolidated the bureaucracy that governed it; they also included measures to foster Saxon mining, such as tax relief for mines experiencing disruptions in productivity, multiple increases in the price of silver, and attenuation of the elector's traditional right to 10 percent of all mining profits. These regulations also organized the purchase of ore by electoral processing and smelting houses. Moreover, like his predecessors in Saxony and princes in the empire more broadly, August also used his influence as elector to encourage new investment in new mining enterprises. In 1579, for instance, he pressured nobles, burghers, city councils, and workers to buy a total of 128 mining shares to set up a consortium to work a new mine (the Tief Marx-Semler-Stollen) in Schneeberg. August devoted attention to other aspects of the mining industry as well. For instance, he built up a workshop in Grünthal that manufactured the sieves, molds, and cables necessary to mine and process ore. His interest in development also extended to the technical details of smelting, particularly of gold and iron ore. In 1556 August not only founded a metallurgical-alchemical laboratory, known as the Goldhaus, behind the electoral palace in Dresden; he drew up the plans for the smelt-ing furnace there himself. Like August's model farms and Anna's botanical gardens, the Goldhaus became yet another site of experimentation and innovation, this time in smelting methods that had direct applications to the Saxon mining industry.[54]

As the remarkably extensive activities and artifacts from August and Anna's court attest, their intellectual curiosity, personal financial interest, and political agenda fostered an intense commitment to innovation in Saxon agriculture and mining. A similarly entrepreneurial outlook can be found at the Stuttgart court of Duke Friedrich of Württemberg. Unlike Duke Julius and Elector August, Friedrich did not inherit this attitude from his ducal predecessors. Because Württemberg was not blessed with the same mineral resources as places like Saxony and the Harz, it lacked the constellation of mining and capital that drew princes in Dresden and Wolfenbüttel into the regional economy. Württemberg, rather, was "a land of small peasant producers," for the most part lacking a *Bürgertum* involved in more commercial economic enterprises. The Stuttgart court, meanwhile, was "narrowly provincial, the territory possessed no style-setting elite." When Friedrich took the throne in Stuttgart, therefore, he inherited a stagnant agricultural economy and low population growth due to plague and crop failure. In response, Duke Friedrich undertook a series of reforms designed to modernize the Württemberg economy.[55]

In 1592, the year before Friedrich began his reign in Stuttgart, the first German translation of Jean Bodin's *Les six livres de la république* appeared in Mömpelgard, where Friedrich already ruled as count.[56] Bodin's articulation of statecraft and absolute sovereignty certainly influenced Duke Friedrich, imparting the lesson that the economy was an important source of princely power. By raising the standard of living of his subjects, Bodin argued, a ruler could increase tax income and thereby strengthen the power of the state. To this end, Friedrich turned his attention to the development of the Württemberg economy, focusing on building infrastructure, encouraging trade, developing new markets, and locating new natural resources. He sought to make the Neckar River navigable by ship between Cannstatt and Heilbronn, for example, thus linking Württemberg to the Rhein River, an important commercial waterway. Friedrich also sought out new mineral resources and invested in the mines in his territories. He established the mining city of Freudenstadt, staffing it with Protestant miners fleeing persecution in Catholic Stiermark in Austria, and bought iron works in Brenztal and Kochertal, placing them under state direction. Duke Friedrich also mined brown coal in Kriegsberg, near Stuttgart, intending it to replace charcoal, which was also becoming a scarce commodity, as a fuel in smelting works. He also forbade the export of raw materials from his territory, displaying an early mercantilist interest in maximizing the use of local materials in production. As Hans-Georg Hofacker put it, "at a downright hectic pace Duke Friedrich sought to implement this early mercantilist economic policy in praxis" by taking on the role of central director of the economy.[57]

Such details of economic policy from across the Holy Roman Empire demonstrate that many of the princely patrons of alchemy in the sixteenth century responded to the general crisis in mining by actively searching for ways to adjust and develop their economies; some of the solutions they devised—such as consolidating state control over mining administration—also dovetailed nicely with a move toward centralization and state building. Natural resources were at the heart of this project. Even Emperor Rudolf II, who is often thought only to have "trod on the paths of secret knowledge with an obsession bordering on madness," lost in his galleries and laboratories, took an interest in practical economic affairs. He issued a number of privileges for the exploration of new ore resources, including one in 1583 for the establishment of a new mining town—Kaiser Rudolfs Stadt (or Rudolfov). Rudolf shared this interest with Vilém Rožmberk, who was the beneficiary of many of these privileges. For Rudolf as well as others, then, developing natural resources—particularly the search for precious metals—often formed a centerpiece of princely political and economic policy in the late sixteenth century, precisely at the same moment that these princes invited alchemists to pursue alchemical projects at court.

The Alchemical Response

As Pamela Long has noted, the "early modern capitalist expansion of mining" created a new audience for technical mining and metallurgical treatises in the second half of the sixteenth century. "Local craft knowledge transmitted orally no longer sufficed for a far-flung group of literate but inexperienced investors," she has argued; therefore, princes and merchant bankers involved in the mining industry extended patronage to "individuals who were able and willing to explain mining and metallurgical practices in writing." One result was a series of treatises in print and manuscript—from the *Bergbüchlein* (ca. 1505–10) of Ulrich Rülein von Calw (d. 1523) to Lazar Ercker's magisterial 1574 *Beschreibung allerfürnehmsten mineralischen Erzt und Bergwerksarten*—that addressed elite audiences who had a financial and political stake in mining. Part of the appeal of these texts was their accessibility and "openness" about metallurgical knowledge. Long has suggested that "the belief in the openness of knowledge and its written transmission, in fact and as an ideal," had significant long-term consequences for scientific authorship in the early modern period. This openness, which became a rhetorical stance of the scientific societies of the seventeenth century, "was an important by-product of the commonality of interest among wealthy investors and the authors of mining and metallurgical books."[58]

Long has juxtaposed sixteenth-century exoteric mining treatises with alchemical writings, arguing that, whereas the authors of mining treatises emphasized openness, alchemical authors emphasized "transmission as an esoteric process, in which an authority transmitted alchemical knowledge to a few initiates usually within an apprenticeship relationship."[59] This important difference in authorial stance is certainly in evidence when comparing printed metallurgical and alchemical treatises; however, it begins to disappear when one's focus shifts from texts to practices, where metallurgical and alchemical operations were closely related, if not indistinguishable. Just as important, the princes and wealthy investors who supported alchemical work in this period clearly saw more commonalities than differences between alchemy and mining; in fact, these patrons clearly thought about alchemy as an extension of their long-standing interest in mining technology. Patrons hired alchemists and mine experts to address the same kinds of technical problems (even if they came up with different solutions), and patrons frequently responded to alchemical proposals with the same kind of investor mentality that framed their responses to mining proposals. This connection between entrepreneurial alchemy and mining would have important consequences for the early modern practice of alchemy, as alchemists were expected to produce not merely ideas, but also increased profits.[60]

The clearest evidence for the perceived continuity of alchemy and mining in late sixteenth-century central Europe comes from the interplay between alchemists' proposals and princes' responses to them. Duke Julius's court in Wolfenbüttel illustrates the close link between alchemy and mining particularly vividly. Julius was interested in alchemy for a variety of reasons, including the medicines, pearls, and gemstones that Philipp Sömmering promised to produce, yet he clearly valued alchemy's contribution to his mining enterprises as well. Commenting on an alchemical proposal in 1576, approximately one year after Sömmering's execution, Duke Julius confessed that he found alchemy to be worth the investment, quite literally: "Like our beloved Lord and Father [Duke Heinrich the Younger] . . . we have been so involved with alchemy that we have given over thousands of thaler to learn it. Nevertheless, it has also taken us so far that . . . we have improved our mines during our reign such that we now enjoy 480,000 gulden more yearly from various mines."[61] Julius's confidence in alchemy's contribution to the ducal mining industries sustained his interest in the art throughout his reign. Despite his negative experience with Philipp Sömmering and his colleagues, Julius credited alchemy with nothing less than the renewal and increased profitability of his mines.

Duke Julius's initial interactions with Philipp Sömmering illustrate these broad connections in more detail. Julius's first contact with Sömmering, in fact, was not about alchemy, but rather in the context of salt mining. The alchemist first came to Wolfenbüttel at the recommendation of Johannes Rhenanus, the mining expert, pastor, and "Salzgraf" whom Duke Julius had hired to consult on the ducal salt mines. Julius took a personal interest in the two saltworks in his territories, Liebenhall bei Salzgitter and his namesake Juliushall bei Bündheim. He refinanced the latter in 1569 and placed it under the supervision of the ducal government in order to make it more profitable. Julius took it upon himself to locate and train skilled workers for the saltworks, and he solicited the expertise of Johannes Rhenanus, who oversaw the productive saltworks in nearby Allendorf an der Werra in Hessen. In September of 1571 Rhenanus arrived in Wolfenbüttel to take a tour of Julius's various mines and make suggestions for how he might make them more productive. He and Julius came to focus on the possibility of conserving wood by switching fuels from charcoal to brown coal, which would make processing more efficient and prevent further deforestation. Sömmering later would take advantage of this concern with fuel conservation as a starting point of a conversation about alchemy.[62]

Rhenanus had first met Sömmering months before he came to Wolfenbüttel in Schmalkalden, where the out-of-work alchemist told Rhenanus about an idea for conserving wood in salt processing and asked the *Bergmeister* to introduce him to Landgrave Wilhelm IV in Hessen. Rhenanus suggested instead that Sömmering try in Wolfenbüttel, since Julius was looking for ways to improve his saltworks in Juliushall.[63] Sömmering took Rhenanus's advice and promptly obtained a position as a salt worker (*Salzsieder*) in Juliushall.[64] The alchemist, however, had loftier goals. He had not come to Braunschweig-Wolfenbüttel merely to work in Julius's salt mines, but clearly hoped to parlay his career as a salt worker into a much more influential position at court. Sömmering later denied that he had arrived with this deliberate strategy, claiming that he "had not seen himself previously as a *Salzsieder*, nor had he presented himself that way; rather, he wanted to conserve wood. It was not his intention, however, to make use of the *Salzsieder* position to make contact with Illustrissimus [Julius]."[65] Whether or not it was intentional, Sömmering's ideas about fuel conservation and salt processing quickly led to an audience with the duke to tell him about an even more valuable secret: alchemy.

When Sömmering finally met Duke Julius, he skillfully crafted his self-presentation to resonate with his desired patron. Aware of Julius's recent introduction of the Reformation into his territories, Sömmering most likely noted his own Lutheran credentials as a pastor ordained by Philipp Melanchthon. The alchemist also relayed the details of his sad fate as a refugee from Gotha, the victim of the political-religious wars of the empire's princes.[66] Having gained Julius's initial sympathy, Sömmering then laid out his full offer. The alchemist predicted that, with his help, "after a time Your Grace will be able to achieve superb yields and profits for this land and its people, particularly with Your Grace's mines, so that you will enjoy two hundred thousand thaler yearly more than before. . . . Moreover, I will share with Your Grace one loth of the Philosophical Tincture by which other lesser metals are made into gold; a principality deserves this, if not more. And then I will disclose and teach Your Grace how to do the process Yourself from beginning to end and thus become the richest and most powerful Potentate in all of Europe."[67] Sömmering's remarkable prediction encapsulated the appeal of practical alchemy for princes like Julius. Sömmering linked alchemy and mining; the alchemist's skill would simultaneously increase the profitability of Julius's mines *and* transmute the base metals in his territories into the more valuable gold. The symbolic profits Sömmering offered, however, were equally important. It was only fitting that a duke should have abundant gold, the alchemist assured his patron, playing on gold's status as a noble metal of unusual purity and permanence. Moreover, Sömmering offered Julius not merely the tincture, but also the recipe; he would teach the duke how to perform a transmutation himself so that Julius could demonstrate not only his ability to create riches, but also his power over nature and, by extension, the human realm as well.

Although Sömmering was not particularly precise about *how* he would apply his alchemical skills to Julius's mines, he probably focused on his knowledge of the separation of metals, or *Scheidekunst*. As veins became less productive in the sixteenth century, methods of extraction and refining became even more important, stimulating an interest in new techniques.[68] Both princes and alchemists recognized that alchemical skills could be particularly useful in this context. For instance, when in 1581 Vilém Rožmberk bought mines in Reichenstein (Złoty Stok) in Lower Silesia, he immediately established an alchemical laboratory there. It is likely that he intended the alchemists to help smelt the gold-bearing arsenic ore in this region, where gold was particularly thinly dispersed and difficult to mine.[69] Similarly, two alchemists proposed *Scheidekünste* to Duke Julius of Braunschweig-Wolfenbüttel shortly after Sömmering and his colleagues' execution. Caspar Uden offered a process "by which copper and silver may be separated," and Theophil Töpfer proposed a somewhat vague process for separating metals "resulting from an alchemical technique."[70] The line between the separation and the transmutation of metals was often quite blurry in the minds of princes and practitioners. One wonders, for instance, what exactly to make of a process that the alchemists Moritz Lam and Georg von Minden offered Duke Julius in 1576, "by which copper can be made out of the Rammelsberg lead."[71] It is unclear whether Julius, who responded positively to the proposal, understood it as a process for the transmutation of lead into copper or as a type of *Saigerprozess* commonly used to separate silver from copper ore using lead. Since Julius's councillor Erasmus Ebener praised the utility of the *Saigerprozess* in the context of the ducal copper mines, Julius probably understood Lam and von Minden's process in this latter light.[72] And yet this was precisely the proposal that caused Julius to exclaim how beneficial alchemy had been to his mines, suggesting that he also regarded the process as alchemical. Clearly princes and alchemists saw a continuity between mining and alchemy, recognizing both as useful techniques for extracting more ore, and wealth, from their territories.

Twenty years later in Stuttgart, the alchemist Georg Honauer offered Duke Friedrich I of Württemberg a process that also demonstrates this link between transmutational alchemy and the separation of metals in the context of mining. Honauer's timing was astute; he arrived just as Duke Friedrich had demonstrated his interest in developing his mining industries by announcing a reward for the location of new ore deposits. The alchemist Honauer claimed that he could produce eight hundred ducats of "fine gold" from one zentner (one hundred pounds) of the iron in Friedrich's iron-rich territory of Mömpelgard. (He could produce even more gold, he added, if the iron was from Hungary.) The alchemist described his process as if it were a typical *Scheidekunst*, or smelting process. "It works like smelting [*wie die Scheiderey*]," Honauer said, "in the same manner as one separates gold and silver, but in this case one separates Mars and Sun [i.e., iron and gold]."[73] Particularly given the large-scale operations Honauer had in mind, his process seems indistinguishable from the myriad techniques for extracting gold or silver from ore.

Everyone involved in Honauer's case, however, regarded him not as a smelter but as an alchemist. Honauer's friend Hans Müller, an imperial goldsmith to Rudolf II, described the process as if it were a transmutation. Müller traced Honauer's process back to an incident several years before the alchemist had come to Stuttgart, when Müller had seen Honauer make gold out of a lead bullet in a military camp in Hungary. "There he began [to practice] that which one calls goldmaking or *Alchymia*," Müller recalled. Müller remembered one incident in particular. Honauer asked onlookers for a lead bullet or cartridge, which could not have been hard to find, given that this demonstration took place on or near a battlefield. While Honauer heated the cartridge over the fire in a cupel, he pulled out a small golden box containing his tincture, which was in a flask specially made to release only a drop at a time, "like a mustard seed." When Honauer added the tincture to the molten bullet, Müller recollected, "it immediately turned into good gold, which was found in an assay to be good." Clearly, in Müller's memory at least, Honauer's art could easily be construed as a transmutation.[74]

The close connection between the separation and the transmutation of metals was the key to metallurgical alchemical work, and it marked the space in which mining entrepreneurs found alchemists to be of the greatest use. From the princes' perspective, the goal was the production of valuable metals; whether this happened by separating base metals from precious metals or transmuting base metals into precious metals made little difference in terms of the outcome, as long as the process made a profit. Thus, whereas in theory natural philosophers, alchemical authors, and probably also most princes recognized that there was a clear difference between *Scheidekünste* and transmutation, in practice, at least in the context of mining, the two commingled much more easily. The frequency with which practitioners, patrons, and other observers made this kind of elision is worth noting, because it reveals the deep connections that patrons could draw between alchemy and mining even as the authors of metallurgical treatises, among others, were trying to sever those connections in print.

Consequences

This connection between alchemy and mining turned out to have important consequences for the practice of alchemy in early modern central Europe. Above all, it made alchemy directly relevant to princes' practical projects and created a pool of patrons willing to include alchemists among the surveyors, mining consultants, and other experts they hired to improve the productivity of their lands. These patrons fueled the demand for alchemy in this period, and their laboratories made large-scale alchemical work possible. By approaching alchemical practice through the lens of their long engagement with mining, however, princes brought a set of expectations about results that had a profound effect on alchemical practice. If late medieval alchemists imagined themselves to be scholars, prophets, and artisans, early modern alchemists like Sömmering found themselves working as entrepreneurs in the grimy world of mining. Technological invention was prized, but so too were profits, efficiency, accessible supplies, and cost; princes made it clear that alchemists' processes were useful only if they were cost effective as well. Alchemists responded accordingly, adopting economic models of productivity as a measure of alchemy's success. Honauer, for instance, took care to point out that each zentner (one hundred pounds) of iron would produce approximately sixteen hundred gulden of gold at a cost of only forty gulden, "and two people could complete one zentner per week." To increase profits, he claimed, Duke Friedrich needed only to employ more workers: "One could also organize it like a large mine," Honauer noted, "with a thousand men."[75] Entrepreneurial alchemists like Honauer were clearly aware of their patrons' concerns about productivity and pitched their projects accordingly.

Negotiations between a man named Kramer and Elector August of Saxony offer a fuller glimpse of princely expectations for alchemical enterprises. Kramer was in fact Heinrich Cramer von Clausburg (or Clausbruch, 1515–99), a tremendously successful Leipzig merchant whose fortune was made trading in fabrics, silks, furs, and gemstones from Antwerp and Cologne to eastern Europe. In the 1560s, Kramer became increasingly involved in mining and metals, investing in a Mansfeld copper mine, lead mines in the Harz, and silver from the Erzgebirge. He had extensive contacts with both Duke Julius and Elector August, facilitating trade in lead and green vitriol (used in dyes and leather production) between Braunschweig-Wolfenbüttel and Saxony throughout the 1560s and 1570s. In 1583, however, Kramer approached August with an alchemical project. Although in the end the bargaining came to nothing, this episode nonetheless suggests that princes thought about alchemical processes in terms of their productive potential and viability as local industries. In this context, transmutation was important not merely for its symbolic value as a demonstration of power or unity, in other words, but as a real technique with potentially valuable economic and political results.[76]

Kramer first contacted Elector August in October 1583 with an offer of the philosophers' stone, but when the stone turned out to be ineffective, August concluded that "it was not yet God's will that such a great secret should be revealed."[77] Shortly thereafter, Kramer tried again to capture the elector's interest by informing him about what he called a "neue Schmelzkunst," or "new art of smelting" that could create gold out of silver. The process worked by dissolving silver in a kind of lye or caustic solution (*Lauge*) until a black sludge flowed from it ("just as one makes copper out of iron with the copper water," he added, referring to a common technique by which pieces of iron were turned into copper by letting them soak in the "copper water"—which contained copper vitriol or copper sulfate—that flowed out of mines). To complete the process, Kramer explained, one then added a final ingredient, an unspecified *composita,* to the mixture, whereupon "it was and remained no longer silver, but rather good, pure gold, just as the copper was made out of the iron."[78] Kramer's description neatly exemplifies the conflation of transmutation and smelting; just as copper water could turn iron into copper, he suggests, so too could his *Lauge,* together with the *composita,* turn silver into gold. Kramer himself did not claim to know how the process worked, but, employing the traditional language of contracts for mining technology, said that he knew of a certain anonymous "inventor" who did, and offered to broker a deal on August's behalf with the inventor's agent.[79] Kramer's use of these terms reveals yet another hidden link between mining and alchemy. Alchemical transmutation, here, was just another "invention" that could be both claimed as a kind of proto-intellectual-property and offered for sale.[80]

The complicated network of people involved in this deal points to a world of enterprising alchemists operating outside (though certainly with an eye toward) the princely courts. Kramer first heard about this process from someone named Marcuß Zellmeyer, who had learned of it from his brother-in-law, Hanß Dreyecker, in Augsburg. At Zellmeyer's behest, Kramer met with Dreyecker to witness the process and evaluate its viability. Kramer's stamp of approval, in turn, encouraged Dreyecker to bring the inventor (who remained anonymous) to Augsburg. Although Dreyecker hoped to sell the process to the Catholic Fuggers for a healthy sum, the inventor refused, as he "was not of a mind to reveal the art to great Potentates, especially not to papists."[81]

At this point Kramer contacted Elector August, because he believed that August would take an interest in the process; moreover, Kramer suspected that the anonymous inventor lived nearby, perhaps was even a subject of Saxony residing somewhere near Leipzig. Kramer swore that he himself did not know the details of the process, however, and that he was not directly connected with the inventor. August was initially suspicious. As his *Kammersecretario* Hans Jenitz (or Jeniss) put it, "but it seemed strange to His Electoral Grace that the inventor demanded such a high price for this art, because if the art was as easy and worked as expeditiously as they said it did, then the demand [of money in exchange for the process] would be excessive and unfair. The inventor could also quickly recover such a sum of money from the art himself, and therefore would need neither an offer nor a contract."[82] Nevertheless, August was intrigued by the process, probably because it offered an opportunity for him to turn the natural resources located in his territory, namely silver, into gold, something even more valuable. He authorized Kramer to bring the inventor to Saxony and to act as the elector's agent in pursuing a deal for the purchase of the alchemical process.

August had a number of reservations about the process that would have to be satisfied first, however. Above all, the Saxon elector was concerned that the process work properly; therefore, he insisted that the gold produced by the process pass a number of different assays.[83] August's interest in the details of the process, moreover, suggests that he was interested in pursuing it as a local industry. He wanted to know what ingredients were necessary to make the *Lauge* that was central to the process. Were they available in Saxony? How long would it take to make the *Lauge*, and how much of it was required to transmute one mark of silver?[84] August was interested not only in whether the process worked, in other words, but also in whether it was efficient and would require importing raw materials from beyond the borders of Saxony. Finally, August wanted to retain exclusive rights to this process and insisted that the inventor promise not to reveal or sell the process to anyone else once the elector had purchased it.[85]

In the end, August backed out of the deal because he thought it might be a scam. He was concerned about the anonymity of the inventor, he told Kramer, and even seemed to think that this inventor did not exist. "Therefore we reject it as fruitless fraud," he wrote, "and we have the impression that the Augsburger [i.e., the inventor's agent] was lacking money and therefore pretended such great things."[86] Despite the negative outcome, however, Elector August's negotiations with Kramer are telling. August expected Kramer's process not only to work, but also to work efficiently, on a large scale, and with native Saxon materials. This transmutation was not merely symbolic, in other words, but it was to be an integral part of August's plan for developing the Saxon economy.

At the end of the seventeenth century, Johann Joachim Becher (1635–82) would develop a sophisticated formulation of the relationship between alchemy and commerce, linking them both through a metaphor of consumption and production. Pamela Smith has argued that, by connecting the two, Becher hoped to translate his noble patrons' traditional interest in alchemy into support for new commercial projects in the Holy Roman Empire.[87] Smith contrasts Becher's seventeenth-century understanding of alchemy as "an activity that mediated between noble and commercial culture" with an earlier view characteristic of courts like Rudolf II's.[88] There, she notes, alchemy was a "cosmic pursuit," in that it not only mediated among religious confessions, but also between the material and the metaphysical.[89] While sixteenth-century alchemists certainly promised precious metals, their operations in the laboratory also served as a "key to nature" and, ultimately, to God. Thus, as Smith argues, "the great work of Paracelsian alchemy as a microcosm of human salvation" in the late sixteenth century began to give way in Becher's hands to "a model of civic *negotium* and manufacture" in the second half of the seventeenth century.[90]

Smith is right that the roots of Becher's commercial alchemy did not lie in the "cosmic" alchemy connected with Rudolf II's court. Instead, to find Becher's predecessors we must look to the entrepreneurial alchemy that princes and bankers pursued in the sixteenth century in connection with increasingly innovative economic projects. In this context, alchemy was but one ingredient in princes' new approach to political economy and statecraft, one that combined traditional ways of generating income for the state (such as mining) with more modern innovations (such as centralized, capitalized economic projects). The Kramer episode, in which a silk and lead merchant from Leipzig nearly convinced the Saxon elector to purchase first the philosophers' stone and then an alchemical process, makes it clear that the princely amalgam of alchemy, commerce, and mining in the sixteenth century was a curious blend of tradition and innovation, indicative of the halting emergence of early modern commerce. For the entrepreneurial alchemists who operated in this context, princely desires for profits and power turned out to be crucial, shaping expectations for their work that often rested uneasily alongside alchemists' abilities and personae.

CHAPTER FOUR

Contracting the Philosophers' Stone

"Two years ago, when I labored on alchemy in various places in my homeland and believed in my foolish mind that I knew a great deal . . . I convinced myself to travel to Your Grace and bring the matter to your attention."[1] So wrote the blind Swiss alchemist Hans Heinrich Nüschler in 1601 when he recalled how he had begun to work for Duke Friedrich of Württemberg. "When I came to Stuttgart," Nüschler continued, "with the best intentions I told Your Grace that I wanted to share the art with you [and that] you should have it tested; only after the assay was found to be legitimate would you present me with gifts as you wished. Before this I desired no money, as this respective contract bears witness."[2] Nüschler and Friedrich did indeed sign a contract, which specified that Nüschler would receive twenty thousand thaler once his process for extracting four loth of gold out of one mark of silver was proven successful. Upon signing, Nüschler gave his patron his alchemical process in writing so that a ducal laboratory worker might try it at Nüschler's expense.

When the process had not yet succeeded after several months, however, Nüschler began to wonder whether in fact his art was as sound as he had convinced Friedrich (and perhaps also himself) it was. More immediately, he worried about the debt he was racking up by financing the duke's tests. Eventually, Nüschler begged Friedrich to halt the tests, but his debt was already more than he could repay. "For more than a quarter year," Nüschler later recalled, "I received nothing, nor at that time did I accept any money. After that, my father often wrote to me furiously that I should come home and take up a proper trade."[3] Unable to fulfill his obligation to the duke or to repay his debts, Nüschler was eventually arrested and convicted of fraud, or *Betrug*.

Nüschler's unhappy story captures many of the mundane aspects of alchemy as it was practiced in the early modern Holy Roman Empire; however pious or scholarly alchemists' goals were, of necessity their practice was often mired in the messy details of patronage, contracts, money, and tests of their abilities. The well-documented princely patronage of alchemy, of course, offered alchemists a range of financial and social benefits. Before alchemists set foot in their patrons' laboratories, however, they had to enter into a whole series of negotiations designed to clarify two lingering problems at the heart of early modern alchemy. First, alchemists had to find ways to establish their own legitimacy. Patrons were justifiably wary of the scores of practitioners who approached them with processes, tinctures, powders, and philosophers' stones. Were these the *Betrüger* against whom Petrarch and Sebastian Brant had warned? Or might these supposed alchemists have very real medical and metallurgical benefits to offer? The fear of false alchemists was amplified by the lack of the licenses, degrees, or guild memberships that offered some sort of credential to practitioners of other trades. Without recourse to such external markers of authority, patrons and alchemists had to invent their own methods of asserting and recognizing alchemical legitimacy; patronage and rigorously controlled demonstrations of skill, therefore, became important tools for determining who was a *Betrüger* and who was not. Even once a patron determined that an alchemist was a legitimate alchemist, however, a second problem remained: what exactly could alchemists be expected to do? Given the multiple understandings of what alchemy was in this period, the answer to this question was not at all self-evident; time frames, costs, ingredients, equipment, expected profits, and so on, all had to be discussed and agreed upon in contracts before alchemical work could begin.

In the late sixteenth century, alchemists and their patrons devised informal means of resolving these two issues. By addressing patrons' concerns about practitioners' legitimacy and establishing some common expectations, these demonstrations and contracts made the entrepreneurial practice of alchemy possible; they did so by creating the conditions in which patrons were willing to invest large sums of money in alchemists and their processes. By formalizing alchemical practices and setting up clear criteria for success and failure, however, these practices also created new risks for alchemical practitioners. If early modern alchemy in general was characterized by a certain fluidity about who could practice it and what they could be expected to do, the practices that appeared around the patronage of entrepreneurial alchemy aimed to pin it down. As Nüschler knew only too well, indeterminacy disappeared when alchemists were asked to put pen to paper and sign a contract promising to complete a process within a certain period of time.

Persona and Patronage

Nüschler's rather simple description of how he came to work for Duke Friedrich of Württemberg masks the complex processes required to find a patron in early modern Europe. Like other kinds of artists and scholars, alchemists often had to enter and navigate the world of the court in order to obtain the rewards of noble patronage. These rewards could be considerable. As a number of scholars have shown, prominent patrons could confer political protection, social recognition, and legitimacy for a whole range of activities that otherwise stood outside (or at the bottom of) social and intellectual hierarchies.[4] Alchemists were in particular need of this kind of legitimation, given their embattled position in the sixteenth century. On a more practical level, moreover, patrons could provide substantial material support. While the wealthiest alchemists, such as Landgrave Moritz of Hessen-Kassel, Count Wolfgang II von Hohenlohe, or Victor August Fugger, had the resources to set up their own alchemical laboratories, most alchemists found it more difficult to fund their own work.[5] Like the late medieval miners who sold *Kuxen* to investors in order to raise money for new mining technologies, therefore, most alchemists were forced to be entrepreneurial, turning to patrons to shoulder the costs of alchemical work in exchange for a share of the profits. Princes and bankers quickly emerged as the most prominent investors in alchemy in the late sixteenth century, just as they had begun to invest in mining companies a century earlier.

Fortunately, the politically fractured Holy Roman Empire provided an abundance of these potential patrons, but alchemists still had to find a way to secure their support. Few practitioners could simply appear at court (as Nüschler apparently did) and expect to have their proposals heard. As Paula Findlen has observed, in fact, "courtly and patrician culture operated on the principle of indirect access" in this period; "if a prince made himself too readily available to his subjects, he diminished his magnificence. Distance as well as intimacy defined one's power."[6] Following this principle, then, many alchemists approached potential patrons indirectly in the guise of the courtier; they operated through social networks, brokers, or carefully crafted letters designed to emphasize not only their own learning and piety, but also the patron's unique suitability to host and support a particular alchemical project.[7] Other alchemists, however, opted to forgo the formalities and rhetorical flourishes; they approached princes directly instead, addressing them as potential investors in the no-nonsense terms of the business proposal. Most alchemists, in other words, had a choice of personae when they approached a prince, and either strategy could be successful, a reminder that alchemists simultaneously inhabited both the world of learning and authorship and the world of entrepreneurship and profits in early modern central Europe.

By controlling access to patrons, brokers and mediators played a crucial role in determining who counted as a legitimate alchemist and who did not. These brokers were not always alchemists themselves, but they tended at least to have some kind of related practical expertise. Philipp Sömmering, for example, first made contact with the Wolfenbüttel court not only through the salt adviser Johannes Rhenanus, but also through Duke Julius's physician, Dr. Jodokus Pellitius, during Pellitius's visit to Julius's salt mines. It was ultimately Pellitius's recommendation that opened the door for Sömmering, who then got the chance to tell Duke Julius about his alchemical projects.[8] In other instances, brokers were alchemists who were already in a particular prince's employ. The alchemist Hans Härpfell, for instance, went to great lengths to bring another alchemist named Wilhelm Haißhammer to the attention of his patron Vilém Rožmberk. Härpfell waited in Prague for months and at great expense for so that he could introduce the Bavarian Haißhammer to Rožmberk's court. Härpfell even had to "borrow Jews' money with a deposit and at high interest" in order to leave something for his wife and children while he was gone in Prague. The sacrifice was evidently worth it to Härpfell, who was certain that Haißhammer's alchemical art would please Rožmberk. As Härpfell assured his patron, the alchemist Haißhammer "worked himself, with his own hands, and his [alchemy] is entirely certain."[9]

As Härpfell's case sugests, this system of patronage appeared to offer benefits not only to alchemists and brokers, but to their patrons as well. Practitioners with the proper social contacts gained a point of entry into the court, a way to transmute social capital into alchemical legitimacy.[10] Brokers, meanwhile, hoped that bringing other talented practitioners to their patron's attention would increase their own status at court. This possibility meant that a strategy of collaboration among alchemists could be preferable to one of denouncing one's rivals at court.[11] Finally, from the patrons' perspective, the patronage system not only facilitated contact between patron and alchemist while maintaining appropriate distance and thus preserving the magnificence of the prince; the very fact that the patronage system linked social capital to alchemical credibility meant that it also served as a filter, ensuring that not all alchemists could gain access to princely patrons.

A surprising number of alchemists broke the unwritten rule of indirect access, however, and approached patrons directly with proposals and projects. Princes well known for their interest in the natural world frequently received letters from projectors hoping to enlist their support. The Saxon elector August and his son Christian I, for instance, received over one thousand pages of letters from various skilled practitioners (*Künstler*) proposing all kinds of projects. In addition to those dealing with alchemy, the projects included techniques (*Künste*) for smelting, making bricks, processing salt, mining, milling, sowing grain, and even constructing an "unheard-of, cruel and secret war weapon."[12] The projects could range into more scholarly topics as well. Some *Künstler* offered to share their knowledge of astronomy and geometry, for example, while one man even offered "a short lesson in what truth and sophistry are, and in theology, astronomy, and medicine."[13] Alchemists and other *Künstler* who chose to write directly to patrons like the electors of Saxony faced a threefold challenge. They had to juggle the multiple tasks of establishing their own credibility, appealing to prospective patrons' learning, and interesting them in particular projects.

Vilém Rožmberk did not receive quite as many overtures as the Saxon electors to the north, but the letters he did collect exemplify the larger patterns among direct patronage appeals. Among the most successful alchemists at the Rožmberk court was the Silesian alchemist Christoff von Hirschenberg. In a 1583 letter to Rožmberk, Hirschenberg spun his curriculum vitae into an elegant tale of the wandering scholar and alchemical adept. After giving an account of his *Wanderjahren* and conversations with scholars, villagers, and artisans, Hirschberg went on to describe how he had then thrown himself into the study of authoritative texts on natural magic and *Chymia*, synthesizing "many wonderful things" and performing a number of experiments. He also offered Rožmberk a tantalizing hint of his approach to alchemy, switching from German to Latin at this point in the letter to describe two different paths to transmutation, the particular and the universal. Hirschenberg hereby pointed to an important distinction that alchemical authors from pseudo-Lull to Robert Boyle made between different types of transmutation. A particular transmutation could occur with any number of oils, powders, and liquids, or *particularia*, but could transmute only a few metals, and even those at only a reasonably low yield (up to five or six times their weight). Universal transmutation, on the other hand, could be carried out only with the philosophers' stone, which could transmute any metal and did so with much higher yields; moreover, the stone itself could be multiplied, yielding potentially limitless transmutations. Obviously the philosophers' stone, capable of universal transmutation, was superior to the *particularia.*[14] Many were familiar with the "particular," Hirschenberg noted, adding that even he could do it if necessary; transmuting with *particularia,* however, was "difficult and laborious." The "universal" transmutation, on the other hand, had the advantage of being "simple," he explained, but it was also "rare," known only to a few. Although many had sought the universal, they had been "deceived" by the many errors in circulation and, most important, by their own ignorance. Not surprisingly, Hirschenberg had more confidence in his own abilities than he did in others'. Because he had built up such a strong foundation of experience as well as both learned and popular knowledge, he assured Rožmberk, God would soon reveal to him knowledge of the "universal," that is, the philosophers' stone. Unfortunately, Hirschenberg continued, years of study had driven him into debt. He now hoped that Rožmberk, as "a particular lover of the liberal arts and studies . . . [of] *Magia* and *Alchimia,*" could provide him with "a comfortable, quiet and safe place" to finish his alchemical work.[15]

Hirschenberg's strategy turned out to be successful because it worked on a number of different levels at once. In this sense, it fits a broad pattern in the "rhetoric of alchemical petitioning" that Bruce Moran has identified at early modern German courts. Moran has noted that alchemists' petitions, like Hirschenberg's, often contained several stock elements: "claims to revelation or to some form of special insight, a rich embroidery of authoritative opinion, personal experience, intrigue, and divine intercession." Moran also points out the centrality of fraud to these petitions, which "inevitably contained condemnations of false philosophers and sophists." Thus, for example, an extremely lengthy petition sent to Landgrave Moritz of Hessen-Kassel by a fencing master named von Stuckhard simultaneously praised Moritz as a learned Christian prince chosen by God to receive the gift of alchemy and condemned the alchemists around him as "poisonous snakes and Basilisks." These elements allowed alchemists, on the one hand, to appeal to their potential patrons as men of learning and faith who deserved to host a successful alchemical enterprise at their courts. On the other hand, such patronage appeals were also an opportunity for alchemists to justify their own knowledge and condemn their rivals. Hirschenberg did so by drawing on the traditional persona of the alchemist as scholar, prophet, and artisan, emphasizing a threefold basis for his knowledge as simultaneously philosophical, divine, and experiential.

At the same time, he echoed printed alchemical texts like Khunrath's and Thurneisser's by claiming the authority of the skilled alchemical expert. By explaining where other "deceived" alchemists went astray in their pursuit of the universal philosophers' stone, he reinforced his own superior knowledge and even suggested that he could serve as a kind of alchemical adviser. This particular rhetorical strategy proved to be quite powerful for many alchemists. In Hirschenberg's case, it led to further negotations with Rožmberk and a contract signed nearly three months later.[16]

Not all alchemists were as rhetorically skilled as Hirschenberg and von Stuckhard, who displayed their skills as courtiers by weaving subtle (or not-so-subtle) tapestries of politics, religion, and knowledge designed to flatter and intrigue potential patrons. Others took a very different approach in their petitions, drawing upon the persona of the mining entrepreneur instead of the courtier to establish their credibility and pique a patron's interest. Tailoring their petitions to princes' specific economic goals, these alchemists chose to emphasize their ability to generate profits rather than their claims to great learning. When the assayer and author Lazar Ercker wrote to Duke Julius of Braunschweig-Wolfenbüttel in 1585, for example, he underscored the money, in the most concrete sense, that he could help generate.[17] Ercker already had connections to the Wolfenbüttel court, for he had worked for Julius's father Heinrich as warden (*Wardein*) of the mint in Goslar from 1558 to 1565; he had also dedicated his treatise on minting, the *Münzbuch,* to Julius in 1563, which earned him a promotion to master of the Goslar mint (*Münzmeister*). In the mid-1560s, Ercker went on to work in Kutná hora (Kuttenberg), Bohemia, where he worked as control assayer (*Gegenprobierer*) along with his wife, Susanne, who served as "manager mistress." After Ercker completed his masterpiece, the *Beschreibung allerfürnemsten mineralischen Ertzt unnd Berckwercksarten,* in 1574, he was promoted to chief inspector of mines (*Oberstbergmeister*) and ennobled in 1586 under the name Ercker von Schrenckenfels.[18]

Ercker pursued various patronage strategies throughout his career, appealing to patrons through both books and profits. As an author, Ercker had access to print as a way of cultivating patronage relationships; his use of dedications, from the *Münzbuch* to the *Beschreibung allerfürnemsten mineralischen Ertzt unnd Berckwercksarten,* which he dedicated to Emperor Maximilian II, shows that he was skilled in the art of rhetorical persuasion. When he wrote to Duke Julius of Braunschweig-Wolfenbüttel in 1585, however, he adopted a very different approach. In this letter, Ercker described a process he had learned from "a Jew from Hanover named Gottschalck" for refining gold gulden, golden coins of lesser quality and purity, and bringing them up to the standard of the ducat, which contained a higher percentage of gold. "By the singular gift of Almighty God," Ercker wrote, "I have invented an art using a powder, by which I can transform Rheinisch or other low quality gold in a few days to proper ducat-quality gold."[19] Ercker claimed that he could refine one hundred marks (about twenty-three kilograms) of Rheinisch gold gulden a week, with an extra cost of ten thaler "for the coals and all the instruments." Moreover, he noted, his process was efficient because he could make use of the by-products as well: "The silver that the Rheinisch gold gulden have in them will be melted out of the powder again," he wrote, "and the gold that the powder has also absorbed, of which there is little, will be separated out as is useful."[20]

As to the profits, Ercker's previous experience led him to predict certain success. "I am of the humble opinion," he wrote, "that for every hundred marks of Rheinisch gold gulden, given the initial costs, there should be a surplus and financial profit of at least seventy or eighty thaler."[21] As support for this claim, Ercker cited his own results using the technique to mint coins for a merchant from Nuremberg. The merchant profited handsomely, according to Ercker, producing in a year as much as two thousand thaler in profits. In Duke Julius's case, Ercker asserted, the profits promised to be even greater. "In my opinion," he wrote to Julius, "it would be much more lucrative and useful to Your Grace because Your Grace can put out [*verlegen*] much better than a merchant—this, however, can not happen at all without this invented art of mine."[22] With this strikingly entrepreneurial language, Ercker demonstrated his awareness of Julius's economic program, framing his alchemical/metallurgical process as a business opportunity, rather than an intellectual project. As these two examples from Ercker and Hirschenberg suggest, alchemists had choices when they appealed directly to patrons. They could position themselves and their patrons as men of learning and piety, as Hirschenberg opted to do, or as entrepreneurs interested in profits, as in Ercker's case. The fact that both strategies could be successful highlights the way in which alchemy could appeal on many different levels at once.

Whether through rhetorical persuasion, social networks, or detailed descriptions of processes and profits, therefore, alchemists tried to overcome fears of fraud and convince patrons of their legitimacy. This barrage of alchemists requesting princely support in the Holy Roman Empire placed political elites in an interesting position as judges of alchemical competence. Some patrons actually sought out assistance with this role by retaining experts at court who served as gatekeepers by evaluating alchemists and their projects. Like brokers, these individuals were only sometimes alchemists but typically had some other relevant expertise. The alchemists Petr Hlavsa and Salome Scheinpflugerin served in this capacity in Vilém Rožmberk's Prague and Třeboň alchemical laboratories, respectively, but so too did Tadeáš Hájek, an astronomer who also served as Rudolf II's court physician in Prague, and Landgrave Moritz of Hessen-Kassel's physician Jacob Mosanus in Kassel.[23] Even if alchemical expertise did not always reside in alchemists themselves, these gatekeepers embodied the persona that alchemists like Thurneisser and Khunrath were carving out in print: the expert who could help sort out the true alchemists from the *Betrüger*.[24] The very existence of such positions, moreover, is an indication of both the fear of fraud and just how many practitioners must have passed through the courts of the Holy Roman Empire. Princes could rely in part on traditional patronage structures to sort them all out, placing their trust in those with the proper social contacts and skills in rhetorical persuasion; but they also resorted to other kinds of screening procedures, putting alchemy to the test in more concrete ways as well.

Demonstrative Proof

Many patrons required demonstrative evidence of alchemical ability, insisting that potential court alchemists (particularly those who promised to make gold or other precious metals) pass a series of tests in order to prove their art. Not unlike artisans, who had to produce a "masterwork" in order to be certified as master artisans, alchemists were often asked to complete a transmutation, after which the resulting gold or silver was sent to the assaying house or another authority on the composition of metals in order to determine whether it was of sufficient quality. When the Moravian Georg Honauer first came to Stuttgart in 1596, for example, Duke Friedrich asked him to perform a series of small tests designed to evaluate the alchemist's claim to make precious metals with a certain powder. Duke Friedrich, in turn, sent several samples of the alchemically produced gold and silver to Dr. Georg Gadner, the duke's chief adviser on mining matters.[25] Gadner assayed the samples according to Friedrich's exacting instructions. What is interesting for the modern observer is that some alchemists' products did pass assays.[26] Gadner found Honauer's alchemical gold to be authentic; according to the results of the first test, the sample was judged to be "at least as good as ducat-quality gold."[27] From the modern perspective, this kind of proof of alchemical transmutation is puzzling indeed—how could experts have certified the alchemical gold as real gold? What are we to make of such stories of alchemical success?

One explanation draws directly on the arguments first articulated by Petrarch and his followers: the assays came up with good quality gold because the sample *did* contain gold, slipped in at some point during the transmutation through sleight of hand.[28] As we have seen, early modern Europeans certainly believed that such chicanery was rampant, and some described these practices in published treatises. Modern historians have often cited these literary accusations as evidence that alchemists actually were committing acts of fraud. On their own, however, such polemical claims simply are not sufficient proof of actual *Betrug*; they must be read as arguments deeply embedded in a battle for alchemical authority, not as transparent documents of fraud. Let us leave aside for a moment the question of whether alchemists *actually* committed the acts of fraud that they were accused of carrying out; we will return to this question by examining the evidence from the alchemical fraud trials in the final chapter. In any case, we need not resort to tricks to explain away the apparent success of transmutations.

One might also blame the apparent authentication of alchemical gold on the tests, arguing that they were not sophisticated enough to determine whether a sample was "really" gold. This explanation is ultimately unsatisfactory as well, however. First of all, as William Newman has shown, the "art-nature debate" had sensitized early modern natural philosophers and alchemists to sophisticated differentiations among different kinds of gold, particularly the much-debated difference between "natural" gold mined from the earth and "artificial" gold produced alchemically. Rather than simply assuming that if something looked like gold, it must be gold, early modern Europeans possessed a nuanced philosophical understanding of the materials they were evaluating.[29] More immediately, textual evidence suggests that since at least the Middle Ages, both alchemists and assayers had utilized highly sophisticated techniques not only for the qualitative and quantitative analysis of metals, but also for the experimental investigation of nature.[30] For example, one medieval text, pseudo-Avicenna's *De anima in arte alkimiae,* described seven tests for assaying gold. As Newman has described them, these tests ranged from visual inspection and taste tests to measurement and chemical analysis of the purported gold. The tests for gold included "attempting its dissolution in 'salts' (if it dissolves it is artificial gold), the use of the touchstone, weight (if the gold is heavier or lighter in specie than normal gold it is fake), loss of its color when fired, ability to sublime, boiling upon fusion, and taste."[31] In the sixteenth century, many of these techniques appeared in print, as the flourishing genre of mining literature increasingly came to focus on assaying techniques. Books such as Biringuccio's *La pirotechnia,* Agricola's *De re metallica,* and, especially, Lazar Ercker's *Beschreibung allerfürnemsten mineralischen Ertzt unnd Berckwercksarten* described numerous methods for assaying the amount of precious metal in a sample.[32] One of the particularly modern characteristics of sixteenth-century assaying, in fact, was a quantitative precision that led one twentieth-century observer to note that the early modern assayer "deserves as much credit as the observational astronomer for providing numerical data and establishing the tradition of accurate measurement without which modern science could not have arisen."[33]

When Duke Friedrich's assayers tested Honauer's gold, in other words, they did more than simply look at it to see if it "looked like gold"; they deployed sophisticated metallurgical expertise. They might have used the touchstone, a technique by which assayers rubbed needles containing a known quantity of precious metal on a dark touchstone; they then compared the color of the resulting streak with the sample. In this case, the evaluation was based primarily on color. Alternatively, assayers might have used the well-known cupellation process.[34] As Biringuccio noted in *La pirotechnia,*

> Cupeling is a very useful thing for anyone who handles gold or silver to know. Indeed it is necessary, because it not only throws light on the work that is to be done, but it also shows the truth. . . . In short, it is the measure by which you have the certainty and safety of knowing that you have not been deceived by art or by your workmen, who had no other interest than their simple wages. Many workmen are found who are so untrustworthy that they no sooner get control of a thing than they think up some fraud. . . . Certainly no mint master, jeweler, or goldbeater can practice his art well without cupellation unless he places his faith in needles and touchstones or in the green patination or other similar shadows of the thing he seeks to know. But the true and surest way is this of the assay.[35]

The assay Birunguccio went on to describe was cupellation, with which any early modern assayer would have been familiar. The process worked as follows: The assayer first melted the sample with lead. A strong coal fire then oxidized the lead, which in turn fused with the impurities or lesser metals in the sample. Once the lead was removed, only a small mass of silver or gold remained behind. By weighing the remaining gold or silver and comparing this amount to the weight of the entire sample, the assayer could determine how pure the sample of alchemical gold actually was.[36]

Friedrich's assayers might also have used another technique involving antimony to assay Honauer's sample; in fact, in this instance, Friedrich specifically requested that "a little of the antimony be used."[37] Biringuccio described this process as well in the fourth book of *La pirotechnia.* In fact, Biringuccio credited alchemists with the discovery of this technique, noting that it "was found by some clever men (alchemists, I believe)." By melting the sample with pieces of brick, antimony, and copper, the gold "will make a residue on account of its heaviness and fall to the bottom as the heaviest thing." The gold thus isolated, the assayer could then skim off the other liquid, heat the remaining gold-antinomy alloy, and oxidize and burn off the antimony, leaving the pure gold behind to be measured.[38] By using such techniques, assayers were able to analyze the composition of "alchemical gold" both experientially and chemically; they could evaluate its purity with remarkable precision.[39]

If polemical treatises are not convincing evidence of the use of sleight of hand, and assaying techniques were good enough to detect samples that did not meet exacting standards of "good, pure gold," then how did alchemists apparently produce precious metals in transmutations? The most fruitful way to understand their success is to try to imagine a worldview in which transmutation *could* occur—a perspective that Bruce Moran has suggested we think of as the "alchemist's reality."[40] Above all, it is important to understand that early modern alchemists and their patrons had good reason to believe that transmutation was possible in theory. Moreover, many claimed to have seen it in practice. Duke Friedrich of Württemberg, for one, was very clear about his views. Despite his experiences with disappointing alchemists like Nüschler, Friedrich was still certain not only that transmutation was possible, but also that he had seen it with his own eyes.[41] Two basic understandings of nature supported this notion. First, a principle that all metals were composed of different proportions and levels of purity of the same two materials (called sulfur and mercury) suggested that, by purifying these ingredients, one could transmute one metal into another. Accompanying this was the idea held by some alchemists that metals naturally strove toward perfection, "ripening" or transforming themselves from base metals into more perfect metals such as gold; the alchemist had only to replicate this natural process.[42] As Moran put it, "from the foundations of accepted religious and philosophical opinion, transmutation was possible and justified. All that was needed was possession of the correct method."[43] The point of the tests, then, was to determine whether the alchemist under examination in fact possessed the correct method.

If alchemical transmutation worked in theory, both Friedrich's claim to have witnessed it and reports of successful transmutations like Honauer's suggest that it also worked in practice (although this is perhaps more difficult for the modern observer to grasp). By translating descriptions of alchemical transmutation into modern chemical terminology, the chemist and historian of alchemy Vladimír Karpenko has dissipated some of the mystery surrounding these processes and made it possible for the modern observer to accept them. Analyzing recipes from the modern chemist's perspective, Karpenko has shown "what really happened" when alchemists appeared to have transmuted base metals into gold, revealing what he has called "the basic techniques used in these alleged transmutations."[44] Karpenko offers the example of the seventeenth-century alchemist Johann Joachim Becher, who proposed a process to extract gold from some silver coins by heating them with a flux. Becher did not use the philosophers' stone or some other kind of agent to effect this transmutation, but instead posited that the fire alone "transmuted" the silver into gold by purifying it.[45] Karpenko explained: "His process consisted of smelting coins (preferably Brabant thalers, which are now known to contain a small amount of gold) with sea-sand and certain salts as a flux. A minute amount of gold was obtained in this way, which Becher mistakenly attributed to the transformation of silver into gold."[46] In a second example, Karpenko describes another typical recipe for transmutation, in which lead was transmuted into gold by means of a golden oil and the action of the fire. Here the golden oil was essential to affect the transmutation. In this case, Karpenko concludes, "the whole process was obviously the reduction of gold which had originally been dissolved in an acid [i.e., the golden oil]. The lemon-yellow colour observed in the first stage of heating could be caused by complex salts of acids of gold."[47] Each of these "alleged transmutations," in other words, is easily comprehensible from our modern perspective. There was apparently always gold in the crucible; the alchemist merely extracted it from an initial material that was not recognized as containing gold (e.g., the coins or the oil).

Karpenko takes pains to point out that his results do not mean that transmutation is possible, but rather that alchemists misread the evidence before them as evidence of transmutation. What we now know to be "the isolation of precious metals from an alloy or from a mixture of compounds," he explains, was thought in early modern Europe to be the creation of gold out of lesser metals.[48] It is certainly satisfying to be able to explain (in modern terms) what alchemists did when they "transmuted" metals. The more relevant question for the historian, however, is not what modern chemistry can teach us about how alchemists "misread" their own processes, but rather what exactly alchemical success meant to early modern Europeans. From the perspective of both alchemists and patrons, these kinds of processes, properly executed, *were* proof of the transmutation of metals. In their view, the reason that there was gold in the crucible at the end of the process was that the alchemist had produced it by means of transmutation (not because it had always been present). Seen from this perspective, alchemical transmutation was possible in practice as well as in theory. In the "alchemist's reality" Honauer *did* produce precious metals alchemically, just as Duke Friedrich's assayers claimed he did; we need not invent a modern explanation simply to explain it away. Tests like these were convincing to audiences that began with the belief that alchemical transmutations were possible, and they worked to assure princes that alchemists who could pass them were qualified, legitimate practitioners of their art. Rather than naively taking alchemists at their word when they promised huge sums of gold, princes thus employed concrete measures, evaluating alchemical transmutations according to the technical standards of the assayer before taking the risk of investing large amounts of money.

Contractual Alchemy

Once a patron was convinced of an alchemist's legitimacy and skill, the two parties often established the alchemical partnership in a formal contract. Early modern Europeans were no strangers to the contract; in fact, even ordinary people would have been likely to sign a number of them during their lifetimes. When parents arranged marriages for their children, for instance, the parties involved would typically set down the terms of the marriage in a notarized marriage contract.[49] Similarly, if they sent their son or daughter to learn a trade in another household, this arrangement, too, might be set down in a legal document.[50] The kinds of documents that alchemists and their patrons signed were part of this broad culture of the contract in early modern Europe, but they were closest to those signed by other skilled practitioners, such as painters, healers, and masons. As a number of scholars have shown, the Renaissance production of art, architecture, and even good health was the product of careful negotiations between patron (or patient) and practitioner.[51] In legally binding contracts, both parties set down their expectations for the relationship in terms of not only pay, but also the quality of the product, the amount of time it would take to complete, and, in some cases, the penalty for failing to complete the project. Such agreements can reveal a great deal about systems of value—whether medical, aesthetic, or intellectual—and of how they collided with economic standards of the marketplace. They can also give us a sense of how early modern alchemists and their patrons understood alchemical success in their own terms.

Like the contracts signed by these other groups, alchemists' contracts were quite specific, stipulating exactly what the alchemist was to accomplish, under what conditions, how long it would take, what the patron was to provide in return, and, in some cases, what was to happen if the alchemist failed. Vilém Rožmberk signed a typical contract on 9 January 1577, with the alchemist Claudius Syrrus Romanus, who had arrived with a recommendation from another Bohemian nobleman and patron of alchemy, Václav Vřesovec.[52] As Vřesovec described him, Syrrus was "so strange and deft, surly, taciturn, that I have not seen anybody equal to him. Yet he is very pious and honest, an absolute master of this Art . . . [and] if I may be allowed to express myself, you should assign to him those two small rooms in the house of Your Honor, that is to say this room and the small closet beside."[53] Evidently Rožmberk was pleased with Syrrus's skill, for Rožmberk and Syrrus then signed the following agreement:

> Agreement proposed to the most illustrious Lord Wilhelm Ursinus of Rosenberg [i.e., Vilém Rožmberk],[54] governor of the house of Rosenberg and the highest burgrave of the Bohemian kingdom.
>
> 1. The soul and body of Claudius will remain free and will belong unto him alone.
> 2. He [Claudius] requests that he will be given protection and support in all matters that are just, legal, and honest.
> 3. No mortal shall enter the living quarters of Claudius during his work, except for the lord, who retains the right to enter where and when he may please.
> 4. His art is to be recognized wherever possible by those unacquainted with it and in ignorance of it. May he be provided with all things necessary for life and work.
> 5. Claudius requests that, if his health is in any way impaired during this work (God grant that this shall not occur), he may receive a suitable pension for the remainder of his days on earth.

6. May the contract between lord [Rožmberk] and Claudius receive the blessing of God. May the medicines made and finished be divided evenly and fairly between lord [Rožmberk] and Claudius.
7. Claudius does not promise anything about the perfection of the magisterium, but he will carry out his art with all his energy and ability; may the Lord God decide about the perfection thereof. If his work may be unsuccessful, Claudius should not be considered an impostor and deceitful. Let the generosity of His Honor provide all else that is needful.

Dated Prague on the 9th day of January 1577.

I, Claudius Syrrus Romanus, sign by my own hand that all that is written above will be carried out with good and holy faithfulness.[55]

In exchange for half of the alchemical medicines Syrrus was to produce, in other words, Rožmberk offered this alchemist full room and board, political protection, and even a sort of health insurance. Moreover, Claudius was sure to make it clear that his success was dependent on God's will alone; he could not be held responsible for failure.

The mutually beneficial nature of Syrrus and Rožmberk's contract indicates that both parties had a hand in crafting this agreement. There is some evidence to suggest that this was more generally the case as well. Duke Friedrich of Württemberg, his envoy in Prague, and the physician Johann Hofrichter, for instance, exchanged several letters over the course of a few months before they set down their final agreement.[56] Finally, responding to Friedrich's draft contract, Hofrichter noted, "I am satisfied with all of the clauses, points, and articles in the contract on which you have worked so hard." He asked for just one more concession: a period of three months in order to move his household to Stuttgart.[57] To the extent that they were mutually negotiated, alchemical contracts were similar to those signed by painters and their patrons, or healers and their patients. As Gianna Pomata has found, in sixteenth- and seventeenth-century Bologna patients and healers both participated in creating contracts that promised the restoration of health in exchange for payment.[58] Michael Baxandall has made a similar point for the contracts that patrons signed with artists in fifteenth-century Italy. He has used these agreements to draw attention to the role of the patron in the production of many Renaissance paintings.[59] Like these others, alchemical contracts were not simple, formulaic documents, but were carefully negotiated by both sides. As Pomata put it, "both voices could be heard" in these agreements.[60]

On the most fundamental level, alchemists' contracts documented an exchange: the patron promised money and support and the alchemist promised a particular process, medicine, or the philosophers' stone. Like Syrrus, most alchemists typically bound themselves in their contracts either to carry out a very specific task for their patron and share the profits, or to teach the patron how to do it him- or herself. Proposals varied widely. Whereas Syrrus, for example, offered Rožmberk half his medicine, the alchemist Daniel Keller promised Marx Fugger one-quarter of the profits from his process for making one ounce of gold out of every mark of silver.[61] Like Keller, many alchemists quantified the yield of their processes precisely, an indication that they knew that their patrons were interested in the details of productivity. Michael Polhaimer, for example, specified that he would make ten loth of silver out of sixty-four loth of mercury, while Hans Hasenbüchler offered to make eight loth of gold out of one mark of silver.[62] It is worth highlighting the specificity of these contracts, which governed only a particular process or product, rather than a more general appointment as court physician, for example. It was not uncommon for court physicians to pursue alchemy as part of their duties at the courts of the Holy Roman Empire in this period; Michael Maier, for instance, carried out his alchemical work in Prague as the personal physician to Emperor Rudolf II.[63] The difference between these two modes of alchemical employment, however, was crucial. For alchemists like Syrrus who signed alchemical contracts, a position at court depended entirely on a single alchemical process; if it failed, then so too did they.[64]

Despite the risks involved, alchemists stood to benefit enormously from these contracts, both in terms of support while carrying out the work and rewards upon completion of the contract. As in Syrrus's case, alchemical employment often included the political protection of the sovereign (*Schutz und Schirm*) for the duration of the contract, not an insignificant benefit given the religious and political turbulence of the Holy Roman Empire.[65] In addition, patrons typically provided the laboratory space, equipment (such as furnaces), and raw materials necessary to carry out processes, as well as a stipend or loan for living expenses while the work was underway. Most contracts also included substantial rewards, to be paid upon the successful completion of the process. Such honorariums were not always monetary. Hans Heinrich Müller, for instance, became Hans Heinrich von Mühlenfels when Emperor Rudolf II ennobled him as a reward in 1603. He later received the village and castle of Neidlingen when he went to work as an alchemist for Duke Friedrich of Württemberg in 1604. Similarly, the prince-bishop Julius Echter von Mespelbrunn promised Conrad von Grumbach the villages of Rimpar and Bergtheim in exchange for his successful process for making gold.[66]

Rewards that did involve money ranged widely. On the low end, Michael Polhaimer was to receive only six hundred gulden from his patron Count Wolfgang II von Hohenlohe for his process of transmuting mercury into silver; at the other extreme, Duke Friedrich of Württemberg promised Johann Hofrichter an astronomical thirty thousand gulden upon completion of the philosophers' stone (in addition to a yearly salary of three hundred gulden while he was working on it). Elector August of Saxony's legal adviser Ulrich Mordeisen offers a rough point of comparison. In 1554, Mordeisen earned a yearly salary of only five hundred gulden, plus a host of other benefits: an additional hundred gulden for expenses, meals at court if he desired, fourteen gulden monthly for the upkeep of his horses, and a yearly delivery of wine, malt for brewing beer, grain, wood, fish, two pigs, one ox, and three barrels of salt pork. Although small in comparison with many alchemists' salaries, Mordeisen's salary was unusually large in order to lure him to August's court and away from his lucrative post as *Ordinarius* at the university law faculty in Leipzig.[67] As the comparison with Mordeisen suggests, alchemists who managed to negotiate court patronage could be very well paid, both while they were working and upon successful completion of the project.

The financial information in alchemical contracts went beyond salaries and bonuses, however; it also reveals the extent to which courtly alchemy was embedded in the same entrepreneurial culture of investment as princely mining enterprises in the sixteenth century. The contracts specified the financial stakes involved, clarifying the relationship between risk and profit in alchemical projects. The contract that Michael Heinrich Wagenmann vom Hoff signed with Duke Friedrich of Württemberg on 23 December 1598, for example, was a remarkably compact document of financial responsibility:

> I, Michael Heinrich Wagenmann vom Hoff, recognize with this, my own handwritten document, that I have received four thousand gulden from the Most Luminous, High-Born Prince and Lord, Lord Friedrich, Duke of Württemberg, and Count of Mumpelgard, and Lord of Haidenhaim, Knight of the Royal Order in France and England. In exchange, I have promised Your Grace—and [I] desire to uphold and carry out [this promise]—to produce the high *Theophrastan medicine,* the universal. And if this does not happen, then so shall I, Wagenmann vom Hoff, be bound to remunerate Your Grace for all working expenses, including

> the four thousand gulden. Together with my wife and child, I am also liable and obliged to bring this to fruition immediately, within ten or twelve weeks, humbly and obediently. I have completed and imprinted this true document with my own noble seal, dated Stuttgart, 23 December, *anno* [15]98.
>
> Michael H. Wagenmann vom Hoff *manu propria*[68]

The contract not only carefully documented how much money was changing hands, but also stipulated that Wagenmann would be fully responsible for all of the expenses incurred should he eventually fail. By attributing financial responsibility for a failed process to the alchemist, contracts like Wagenmann's minimized the patron's financial risk in supporting the project.

The temptation is great to see these contracts as foolish on the part of patrons; it might seem that, like Erasmus's character Balbinus, princes let their hope for a miraculous source of gold or health corrupt their good sense and convince them to pour money into the pockets of deceitful alchemists.[69] Clauses that placed the financial burden on alchemists, however, show such contracts to have been rather savvy deals from the patrons' perspective. Wealthy investors simply had to provide the capital and resources to finance alchemical projects. Should a process fail, contracts could ensure that the debt fell to the alchemist; should the alchemist succeed, however, the profits, at least in part, went to the prince. From this perspective, contracts actually seem like a fairly foolproof way to avoid Balbinus's fate as a victim of *Betrug*. They created legally binding obligations that saddled the alchemist, rather than the prince, with most of the risk. Furthermore, it is important to remember that princes entered into these arrangements with the reasonable expectation that the alchemists would indeed complete their work as stipulated by the contract—otherwise they would not have signed the agreements. Assuming that alchemical processes could be successful, the princes stood to reap substantial profits from their initial investments. Jost Weyer calculated, for instance, that the augmentation process that Michael Polhaimer contracted to carry out for Count Wolfgang von Hohenlohe would have doubled Wolfgang's initial investment.[70] Additional costs aside, Polhaimer and Wolfgang expected to make a healthy profit. Alchemical contracts thus had much to offer wealthy patrons, allowing them to transfer the bulk of the risk to the alchemists but retain profits for themselves.

But why did alchemists enter into these deals, given that they risked not only the anger of a disappointed (and potentially very powerful) patron, but also the incursion of huge debts? First, contracts offered alchemists distinct financial advantages. Because, like mining, large-scale alchemy was so capital intensive, such contracts were one of the few ways that ordinary alchemists could secure the resources needed to carry out their work. Today we tend to imagine alchemists as Midas figures who could produce instantly huge quantities of gold, yet we tend to forget about the considerable sums of money and raw materials alchemy often required. The two recipes for the philosophers' stone that Anna Zieglerin related to Duke Julius in Wolfenbüttel in 1573, for instance, required six pounds of lead, two large rubies, one small bird, eight pounds of quicksilver, vinegar, salt, horse manure, and unspecified quantities of Hungarian gold, antimony, and the quintessence. Once the stone itself was finished, of course, Julius would need additional supplies to use it in the production of medicines, precious metals, and stones.[71] Alchemists working on a larger scale had even more substantial and costly supply lists. When Georg Honauer began his work for Duke Friedrich of Württemberg in 1597 (after passing the initial trial of his skill), he placed the following order for supplies:

From the armory:
10 zentner [468 kg/1030 lbs] saltpeter
18 zentner [842 kg/1852 lbs] lead
as much vitriol [*Kupferwasser*][72] as possible
7 zentner [327 kg/719 lbs] iron [*Schöneisen*]
From the palace kitchen:
3 zentner [140 kg/308 lbs] salt
From Nuremberg:
20 zentner [935 kg/2057 lbs] white copper [*Kupferweiß*]
10 zentner [468 kg/1030 lbs] mountain antimony
4 zentner [187 kg/411 lbs] glass gall [*Glasgalle,* a salt skimmed from molten glass]
2 zentner [94 kg/207 lbs] Venetian soap [*venedische Seife*][73]

As supply lists like this suggest, alchemy could be a truly large-scale and capital-intensive enterprise. On their own, few alchemists could afford or had access to these materials and the ovens, cupels, and other equipment necessary to work them; in exchange for a share of the resulting profits, however, wealthy patrons would finance this alchemical work. Thus, by providing the supplies alchemists needed to pursue their art at court, contracts worked very much to alchemists' advantage as well as princes'.

The benefits of contracts went beyond funding, however; these agreements offered alchemists legitimacy as well. On the individual level, alchemists could use the very fact that they had secured a contract with a prince as a kind of certification of their skill. Remarkably, Syrrus's contract stated this explicitly, stipulating, "His art is to be recognized wherever possible by those unacquainted with it and in ignorance of it."[74] In effect, Syrrus's contract claimed that the prince's recognition of his skill should lend him credibility as an alchemist in the eyes of those who had not personally observed his work; the contract was to function as a de facto license. Even when contracts did not claim this authority so explicitly, still they could be seen as a stamp of approval, a legitimization of a particular individual alchemist's claims in a chaotic but competitive marketplace. A contract was evidence that to some extent an alchemist had proven his or her claims and managed to convince a patron that the purported art was worth gathering the resources to pursue further. For alchemists struggling to assert the persona of the legitimate alchemist in the face of literary and artistic ridicule, this kind of princely legitimation could be crucial. Alchemists could point to their contracts as proof that they were not to be confused with the hoards of *Betrüger,* but rather to be "recognized wherever possible" as genuine practitioners who had withstood rigorous screening procedures at court. In this sense, contracts not only documented an exchange of goods and services; they could also serve as a source of authority and legitimacy for alchemical practitioners.

For both alchemists and patrons, then, the contracts offered advantages on several levels. From the patron's perspective, contracts constructed legally binding financial arrangements that minimized risk while securing access to profits. By transferring patrons' risk to alchemists and raising the stakes of pursuing patronage, contracts offered patrons a further advantage. Many alchemists must not have been willing to accept that level of risk; alchemists who were, however, stood to gain not only absolutely crucial material support for their work, but also much-needed credibility for their claims to be legitimate alchemists, as opposed to *Betrüger.* By simultaneously addressing patrons' and practitioners' multiple concerns, alchemical contracts played a central role in facilitating alchemical practice as not just private study, but as a collaborative, entrepreneurial endeavor.

Finally, alchemical contracts served as moments of definition and decision in an otherwise indeterminate alchemical landscape. Given the fluidity about ideas such as who should practice alchemy, what its goals should be, whether it was best situated as a part of natural philosophy, medicine, metallurgy, or all three, and so on, the contract was a useful moment for both parties to make sure they agreed about what it was they were doing, in the specific instance, at least. How long would the medicine or tincture take to make? What ingredients did it require? Would the process work automatically, or was it dependent on God's favor? The details of a contract answered these questions, becoming a statement of what was required to complete a particular alchemical process successfully. This moment of definition was crucial because it provided a baseline from which patrons could later evaluate whether or not an alchemist had accomplished his (or her) goals. By clarifying up front, for example, how long a process should take, exactly how much it would cost, what the yield would be, and so on, princes entered into agreements with enough knowledge to later evaluate alchemists' achievements. Any additional request for more time or materials, moreover, became in effect a supplication for special favor, for extraordinary indulgence on the prince's part in a failing process. In this sense, each contract set down anew a working definition of alchemical success. Contracts were, in essence, a way for both parties to agree on what alchemy *was*—not universally, perhaps, but at least so that a particular patron and an alchemist could work together to put alchemy into practice.

In this respect, alchemists' contracts paralleled the cure contracts that Gianna Pomata has examined in Bologna. Like the patrons and alchemists who spelled out precisely what would count as an alchemical success, healers in early modern Bologna articulated exactly what they meant when they said they would make a patient healthy. Rogerio de Bruch of Bergamo, for instance, assured Bosso the wool carder that, after treatment, "you will be able to feed yourself with your hand and cut bread and wear shoes and walk and speak much better than you do now." Rogerio promised, "I shall take care of all the expenses that will be necessary for this; and at that time, you shall pay me seven Genovese lire." Bosso the wool carder had to do his part as well. As the contract continued, he promised "not [to] eat any fruit, beef, pasta—whether boiled or dry—or cabbage."[75] Contracts like these served as an agreement between patient and healer about what both parties should expect. In this case, if Bosso were not able to cut bread after his cure, he would not have to pay Rogerio de Bruch for his services since, as the contract stipulated, the cure would not have worked. Similarly, if Heinrich Wagenmann vom Hoff had not produced the Theophrastian universal "within ten or twelve weeks," as promised, he would have been bound to remunerate his patron Friedrich for all expenses incurred.

In other respects, however, cure contracts varied significantly from their alchemical counterparts. Pomata has drawn attention in particular to the power dynamic involved in cure contracts. As documents crafted by both parties in the sixteenth century, she argues, "these agreements represented a compromise between the interests of the healer and those of the patient. Their aim was to redress the imbalance in the patient-healer relationship which resulted from the healer's superior knowledge and expertise."[76] Alchemists, too, had specialized expertise and participated in setting the terms of their agreements. But as sovereign rulers as well as prominent patrons of alchemy, princes held immense power that outweighed any most alchemists might possess. Thus, whereas cure contracts "served as an instrument to minimize the patient's disadvantage," alchemists' contracts worked to the contrary: to minimize the *alchemist's* disadvantage in contracting with powerful patrons.

In many ways, contracts were crucial to putting entrepreneurial alchemy into practice in early modern central Europe. They addressed the concerns of both alchemists and patrons, from financial responsibility to the more intangible question of legitimacy. In this sense, both parties stood to gain from these arrangements. Nevertheless, at the heart of early modern alchemical contracts lay a vast power and risk differential between patron and alchemist that would have enormous consequences. The unimaginable rewards that awaited successful alchemists set them apart from other kinds of practitioners like artists or miners. But then again, the rewards were high because so were the risks. Alchemists who could not meet their contractual obligations risked major debt or the possibility of breaking a contract with a powerful patron.

In the end, the solutions that alchemists and their patrons devised to the indeterminate status of alchemy may have created as many problems as they solved, for practical alchemy probably worked best for all concerned when expectations on the part of both parties were left vague. When patron and practitioner were forced to express their expectations in written contracts stipulating precisely what the alchemist would produce and when, they increased the likelihood that they would fail to meet those exacting expectations and opened alchemists up to fraud charges. Ultimately, situating alchemy in the framework of the mining industry was both a boon and a curse for entrepreneurial alchemists. It provided the possibility of employment with patrons who could confer both wealth and status. But by operating within the culture of entrepreneurship and investment, alchemists and patrons alike increasingly judged alchemical success by standards they could not meet. For alchemists who failed to fulfill their contracts, as Nüschler and others learned, the early modern penal system awaited.

CHAPTER FIVE

Laboratories, Space, and Secrecy

What sorts of spaces did alchemists use to carry out their work in the cities and courts of the Holy Roman Empire? All alchemists, however philosophical their goals ultimately were, required some kind of laboratory in which to perform operations at the fire. But what did those laboratories look like? Could practitioners set up a laboratory anywhere with a fire, or did alchemical work require a specific kind of space? Furthermore, how important was secrecy? Given its importance in the alchemical textual tradition, we might imagine that secrecy was paramount for alchemical practitioners. "The first precept is that the worker in this art must be silent and secretive and reveal his secret to no one," the author of the medieval *Libellus de Alchimia* warned; ". . . he should have a place and a special house, hidden from men, in which there are two or three rooms in which are carried out the processes for sublimating and for making solutions and distillations."[1] This tradition remained powerful, but early modern alchemists also had other reasons to be concerned about secrecy; as one seventeenth-century author offered, "if one of the rulers should learn about such a person [who has successfully completed the philosophers' stone], he would want to seize him and make him a virtual prisoner for life. Therefore they must keep silent."[2] Should the laboratory, then, be isolated, situated somewhere beyond the prying eyes of public and princes alike?

These kinds of questions about the distinctiveness and physical location of workspaces were a crucial part of putting alchemy into practice in the early modern period. A makeshift laboratory smack in the middle of an urban kitchen, as Brueghel imagined it in his mid-sixteenth-century engraving, would permeate alchemy with a very different set of meanings and practices than a laboratory hidden away inside a princely palace. Early modern alchemists themselves understood this too, as the well-known example of Andreas Libavius's Chemical House makes clear. The plan for the imagined Chemical House, which the German schoolmaster included in the 1606 commentary on his 1597 textbook, *Alchemia*, was designed as a critique of the Danish astronomer Tycho Brahe's laboratory at his observatory and research center, Uraniborg. To Libavius, Tycho's alchemical laboratory, tucked away on the island of Hven, exemplified "*Burgwissenschaft*," or "castle-science," characterized by secretiveness, aristocratic aloofness, and dedication to the contemplative life. By modeling his own Chemical House on an urban *bürgerlich* house, in contrast, Libavius sought to symbolize his belief in the openness and civic engagement of his art. Despite these gestures toward openness, however, Libavius's design did not abandon entirely the traditional secrecy associated with alchemy. Common *Laboranten* (laboratory workers) in the Chemical House were to perform more mundane alchemical processes in relatively public spaces, but the production of that *arcana arcanissima*, the philosophers' stone, was still to be completed in the most private and sequestered of spaces, away from the crowd. With this physical separation, Libavius claimed that the most important alchemical tasks were best left to a private space belonging to the master alchemist alone; only the privileged were to have access to alchemy's greatest secrets. By articulating alchemical principles architecturally, Libavius demonstrated his awareness that decisions about the location of laboratories and the partition of space within them were central to defining alchemy's place in early modern Europe.[3]

Libavius's Chemical House has become emblematic of the way in which space expressed the ideal purposes and practices of knowledge in the early modern period.[4] The Chemical House, however, remained imaginary; although it reveals a great deal about Libavius's views of alchemy, it tells us little about where alchemists actually carried out their work. The same is true for the rich and abundant paintings from the Netherlands that, following Brueghel, took up the subject of the alchemist in his laboratory in the seventeenth century. These images are a fantastic source for understanding the cultural appeal of the theme for urban viewers. The images elaborated visually on humanist discussions of the alchemist as a moral subject, and they conveyed a range of positions on the subject. Moreover, they put alchemy on display, offering up the elaborate apparatus of the alchemical laboratory in remarkable detail for curious viewers. As much as they tantalize us with their scattered equipment and operations in progress, however, none of these early modern laboratory images was meant to be a transparent reproduction of a real alchemical laboratory, nor can they be read as such. Like alchemical personae discussed in chapter 2, these early modern representations of alchemical laboratories should be read as arguments, as "iconographically rhetorical" representations of what their authors imagined or hoped alchemy to be, not as realistic renderings of the spaces in which alchemists actually worked.[5]

For some insight into real, brick-and-mortar alchemical laboratories, historians of alchemy have turned to the archives and to archaeology. Both the archaeological work on a laboratory in Oberstockstall, Austria, and Jost Weyer's study of Count Wolfgang II von Hohenlohe's laboratory in Weikersheim have revealed a great deal about the material culture of two important sixteenth-century laboratories.[6] More recently, William Newman and Lawrence Principe's creative and meticulous study of the laboratory notebooks of the seventeenth-century chymist George Starkey (alias Eireneus Philalethes) has revealed "what 'Philalethes' was doing from day to day in his laboratory, how he thought about his laboratory activities, and how he set about designing, executing, and evaluating them." Starkey's extraordinary notebooks offer rare insight not only into the practices of an influential seventeenth-century chymist, but also into the gaps between his private practice and the way in which he chose to represent that work publicly in his (or rather Philalethes's) printed texts. We still have a great deal to learn about working alchemical laboratories, however. These studies have contributed crucial information about the kinds of apparatus one might find in an early modern alchemical laboratory, as well as the role of laboratory work in the construction of alchemical/chymical knowledge. Nonetheless, these studies have not addressed in detail the use of space in organizing early modern alchemical work. As Newman and Principe have pointed out recently, "the reality of a working alchemical laboratory remains somewhat elusive."[7]

And yet alchemical laboratories proliferated in the early modern period as alchemy's visibility rose. In the cities, courts, and cloisters of the Holy Roman Empire, practitioners and patrons with the means to do so built special buildings for their alchemical work, while others improvised with whatever spaces were available: kitchens, churches, apothecary shops, and workshops. Archival remnants of some of these spaces do, in fact, remain; these bits and pieces—inventories, architectural details, supply orders, and reports—offer a glimpse of their contours and how space organized the activity inside that can complement the work of archaeologists. This view from the laboratory floor, as it were, not only can begin to fill in some very basic details about how space was employed to organize the production of alchemical knowledge; it also confirms the importance of some of the main themes raised by the more idealized representations produced by people like Libavius, Brueghel, and his Netherlandish followers. The designers and denizens of laboratories, in fact, were focused on many of the same questions raised by imaginary laboratories, particularly the relationship of alchemy to other related arts and the problematic role of secrecy. This shared set of questions did not produce the same answers, however. As concerned with these issues as alchemists and their patrons were, as we shall see, the laboratories they actually built resolved such questions very differently from the ways in which their imaginary counterparts did.

Duke Friedrich I's Laboratory in Stuttgart

Let us begin our foray into early modern alchemical laboratories with Duke Friedrich's Alte Lusthaus (Old Summer House) in Stuttgart. Friedrich's laboratory came into existence in 1596, when he informed his *Baumeister* (chief architect) Heinrich Schickardt and his alchemical assistant Lucas Osiander the Younger that the Alte Lusthaus in the ducal garden was to be converted into an alchemical laboratory. Originally built in 1553 under Duke Christoph, by Friedrich's reign the Alte Lusthaus was situated in what he described as the zoological garden (*Thiergarten*), between what were eventually to become a racetrack and a labyrinth.[8] Superseded by the more spectacular Neue Lusthaus (New Summer House) in 1593, the Alte Lusthaus was available when Georg Honauer appeared at Friedrich's court three years later to transmute large quantities of iron into gold.[9] Friedrich converted the space into a laboratory and used it as his central alchemical laboratory until his death in 1608. A 1616 engraving from Matthaeus Merian shows the Alte Lusthaus (no. 6) and its location in the ducal gardens in Stuttgart (fig. 7).[10] The much grander Neue Lusthaus that replaced it is visible on the right (no. 5); the characteristically Renaissance colonnade and towers of the more recent Neue Lusthaus distinguished it from the more medieval-looking Alte Lusthaus.[11] Aside from its availability, several features of the Alte Lusthaus's location would have made it desirable for an alchemical laboratory: its proximity to the ducal palace (no. 1) allowed Friedrich to keep an eye on his alchemists, while the laboratory's isolation out in the garden kept the stench and smoke away from the court.

We might imagine Duke Friedrich strolling across the gardens to check on the progress of his alchemists in the Alte Lusthaus.[12] Stepping out from his palace (no. 1), past the menagerie on the right and through the western

Figure 7. Engraving of Stuttgart ducal gardens by Matthaeus Merian, in Esaias van Hulsen, *Repraesentatio der fürstlichen aufzug und Ritterspil* ([Stuttgart?], 1616). (Beinecke Rare Book and Manuscript Library, Yale University.)

gate to his ducal gardens, he might have noted the kitchen to his left as he walked eastward past the gate to another garden, the water mill and on toward the Alte Lusthaus. The two-story building was not nearly as imposing as the Neue Lusthaus; it was approximately one hundred feet long and forty-four feet wide. Duke Friedrich would have entered the building through a stairwell on the north side of the building. He might have bypassed the ground floor, which was divided up into eight smaller rooms, two hallways, and one large hall or room; this floor was perhaps where the *Laboranten* resided or stored supplies. The alchemical work most likely took place on the upper floor, which contained one large hall and two smaller rooms. The building had a two-story attic as well, which is where all of the alchemical equipment ended up when, decades later, the building was converted into the ducal *Kunstkammer.* According to a 1634 inventory, the attic then contained "many large and small glass cucurbits and other glasses, as well as many different things belonging to the laboratory." Signaling a shift in ducal priorities by the mid-seventeenth century, the officials compiling the 1634 inventory added that this alchemical equipment "would take much time to describe, but would not be worth the trouble."[13]

Standing in the upper rooms of the Alte Lusthaus, Duke Friedrich would have had a pleasant view of his gardens through the numerous windows. The large hall was certainly well lit by day, as it contained a total of fourteen windows (including those in the corner niches) and also received light from the hexagonal stairwell. Of the two smaller rooms, the northeastern room contained two windows as well as three in the corner niches, while the southeastern room contained only one window plus the three in the corner niches. This was hardly the dim, underground laboratory that critics imagined alchemical workspaces to be.[14]

The fire was the heart of all alchemical work and was thus the natural center of any laboratory. The Alte Lusthaus in Stuttgart had an *Esse,* or an open fireplace with a chimney of the sort used by smiths to heat iron until it glowed and became malleable. Alchemists typically used this kind of hearth for manipulating a variety of materials, or they placed smaller portable furnaces inside it so that the chimney would draw the poisonous vapors out of the room.[15] Friedrich's alchemists and *Laboranten* certainly used such small furnaces, whether in or out of the hearth, as a 1608 inventory of the laboratory indicates that it was equipped with fourteen portable "copper furnaces of various types" and four "copper ovens large and small."[16] Although we do not know more precisely what kinds of furnaces or ovens these were, some of them were almost certainly distillation furnaces of the sort depicted on the title page of Lazar Ercker's *Beschreibung sllerfürnemsten mineralischen Ertzt unnd Berckwercksarten.* There were probably also assaying furnaces, into which the material to be melted was placed through an opening in the front of the furnace. There may also have been other standard sorts of furnaces, including the common *Verbundofen.* This furnace was also known as a "Faule Heinz" (Lazy Heinz) because it allowed the practitioner to add a large amount of fuel in a central chamber, which then slowly dropped down to side chambers as necessary—a particularly useful furnace for long operations that required feeding the fire constantly (fig. 3).[17]

Looking around the large hall, Friedrich would have seen a variety of glass distillation vessels. Typically, these included three parts: the material to be distilled was heated in the bottom vessel, or cucurbit, condensed in the alembic (distillation head, or *Helm*) on top, where it then cooled and dripped through a "beak" into a receiver (*Recipient*). (See the *Laborant* in the center of fig. 3.) Friedrich's *Laboranten* had a small "separation cucurbit," for instance, which could have been used for separating silver and gold with parting acid.[18] Friedrich also would have seen two "pelicans" (one of which was "a lovely pelican . . . brought here from Venice"), a type of circulation vessel in which liquid was heated into a vapor, condensed in the top portion of the vessel, and then returned to the bottom of the vessel via two side tubes. The name "pelican" drew on medieval bestiary lore that the pelican fed its young with blood from its own breast, much as Christ had sacrificed his blood for humanity (fig. 8).[19] In addition, Friedrich would have seen fourteen copper *balnea*—baths for gently heating and purifying materials in water, sand, or ash rather than placing them directly in the fire—with all of the accompanying rings and racks to keep the vessels in place.[20]

Assaying equipment was essential to any alchemical laboratory, and Duke Friedrich's laboratory was well equipped to carry out such fundamental operations.[21] It contained about one hundred different ceramic vessels for manipulating metal samples in the fire, including some bone ash cupels for use in purifying silver, as well as brass press molds for making new ones (fig. 9).[22] Because alchemists and *Laboranten* needed a range of balances to weigh ingredients and measure their results, the laboratory was also outfitted with "a box with two assaying balances, . . . the accompanying assaying weights for all kinds of ore, . . . an assaying balance in a small container from Müllenfels,[23] . . . a little balance with a half-pound weight, eight balances with large and small brass pans, six weights from eight pounds to half a pound, a large balance with iron pans[, and] weights from a zentner to one pound" (fig. 10).[24] Friedrich's alchemists probably also determined the amounts of gold or silver in samples (albeit more provisionally) by rubbing the unknown sample on a touchstone, then comparing the streak with a silver or gold assaying needle of known composition. Their laboratory was supplied with several different types of touchstones and assaying needles for this purpose (fig. 11).[25]

Finally, in surveying his laboratory Friedrich certainly would have observed all sorts of minerals and metals used in alchemical processes. At Friedrich's death in 1608, for instance, the laboratory was well stocked with over one hundred pounds of antimony for separating silver and gold and nearly two hundred pounds of white and yellow arsenic, which was particularly valued for its ability to "whiten" copper.[26] In addition, the laboratory stored lead, "ruddy stone" (probably *lapis haematitis*, known today as a naturally occurring iron oxide), salts, borax, vitriol, and "a mixture of lead and antinomy made by Brunhoffer," or Georg Honauer (allegedly Herr zu Brunhoff), who had arrived in Stuttgart in 1596.[27] This laboratory was certainly well stocked with the mineral resources of the Württemberg state.

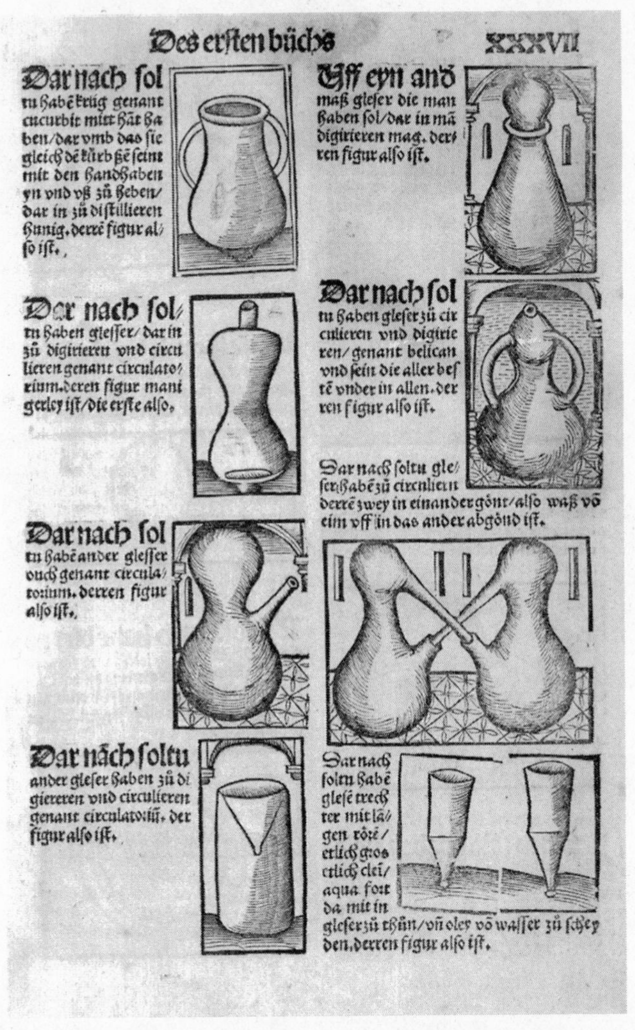

Des erſten büchs XXXVII

Dar nach ſol tu habē krüg genant cucurbit mitt hät haben/dar vmb das ſie gleich dē kürbſē ſeint mit den handhaben yn vnd vß zů heben/dar in zů diſtillieren hunig. derrē figur alſo iſt.

Dar nach ſol tu haben gleſſer/dar in zů digirieren vnd circulieren genant circulatorium. deren figur manigerley iſt/die erſte alſo.

Dar nach ſol tu habē ander gleſſer ouch genant circulatorium. derren figur alſo iſt.

Dar nách ſoltu ander gleſer haben zů digiereren vnd circulieren genant circulatorium. der figur alſo iſt.

Vff eyn andˀ maß gleſer die man haben ſol/dar in mā digirieren mag. derren figur alſo iſt.

Dar nach ſol tu haben gleſer zů circulieren vnd digirieren/genant belican vnd ſein die aller beſtē vnder in allen. derren figur alſo iſt.

Dar nach ſoltu gleſer habē zů circulieren derrē zwey in einander gönt/alſo waß vō eim vff in das ander abgönd iſt.

Dar nach ſoltu habē gleſē trechter mit lāgen rörē/etlich gros etlich clei/aqua fort da mit in gleſer zů thůn/vn̄ oley vō waſſer zů ſcheyden. derren figur alſo iſt.

Figure 8. Pelican (*far right, second from top*) and other glassware, from Hieronymus Brunschwig, *Liber de arte Distillandi* (Strasbourg: Johann Grünignger, 1512). (From the Roy G. Neville Historical Chemical Library, a collection in the Othmer Library, Chemical Heritage Foundation. Photo by Douglas A. Lockard.)

Contrary to the stereotypes, then, the Stuttgart laboratory was not hidden underground, bathed in darkness and secrecy, nor was it crammed into a corner of the palace kitchen.[28] Rather, the laboratory was founded in its own distinct space in the midst of the ducal gardens, a short stretch from the marvelous Neue Lusthaus, which surely captivated the attention of

Figure 9. Cupels, molds, and balls of ash, from Lazarus Ercker, *Beschreibung allerfürnemisten mineralischen Ertzt unnd Berckwercksarten* (Frankfurt am Main: Georg Schwartz, 1580). (From the Roy G. Neville Historical Chemical Library, a collection in the Othmer Library, Chemical Heritage Foundation. Photo by Douglas A. Lockard.)

the duke and his courtiers. The rooms were well ventilated, bathed with light, and certainly at least as full of distilling vessels, assaying balances, and furnaces as any engraving or painting of an alchemist. And, most important, this laboratory was bustling. Standing in the midst of his primary laboratory, Friedrich would have seen several of his alchemists and

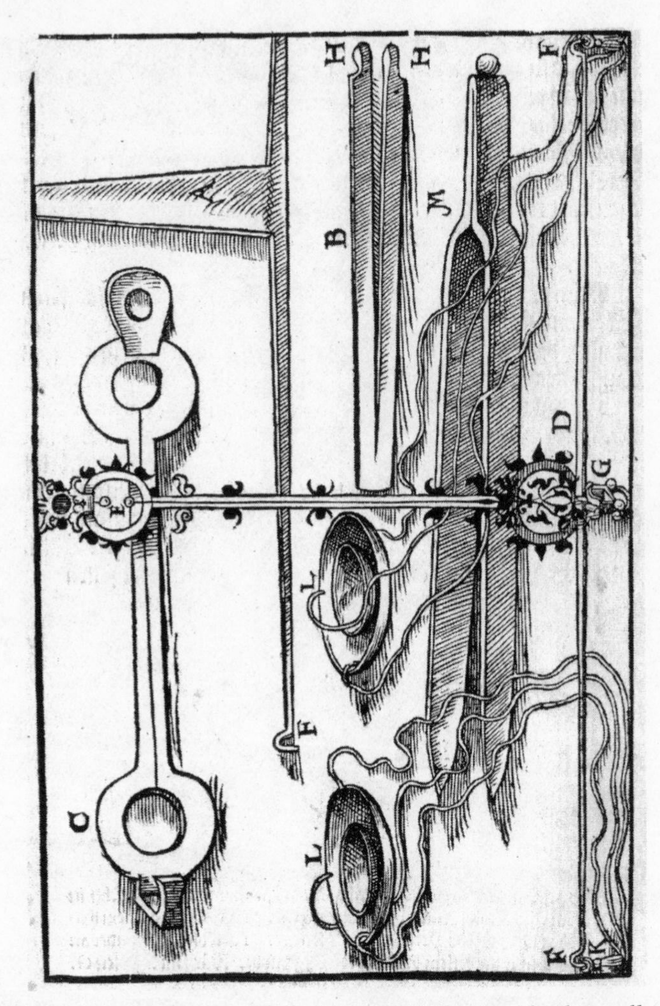

Figure 10. An assay balance in parts, from Lazarus Ercker, *Beschreibung allerfürnemisten mineralischen Ertzt unnd Berckwercksarten* (Frankfurt am Main: Georg Schwartz, 1580). (From the Roy G. Neville Historical Chemical Library, a collection in the Othmer Library, Chemical Heritage Foundation. Photo by Douglas A. Lockard.)

Laboranten at work at once, sharing equipment, furnaces, and presumably information about their work.

Alchemist, *Laborant*, and *Inspektor*

Fortunately, the same inventory that recorded the contents of Friedrich's Stuttgart alchemical laboratories also documented the activities carried

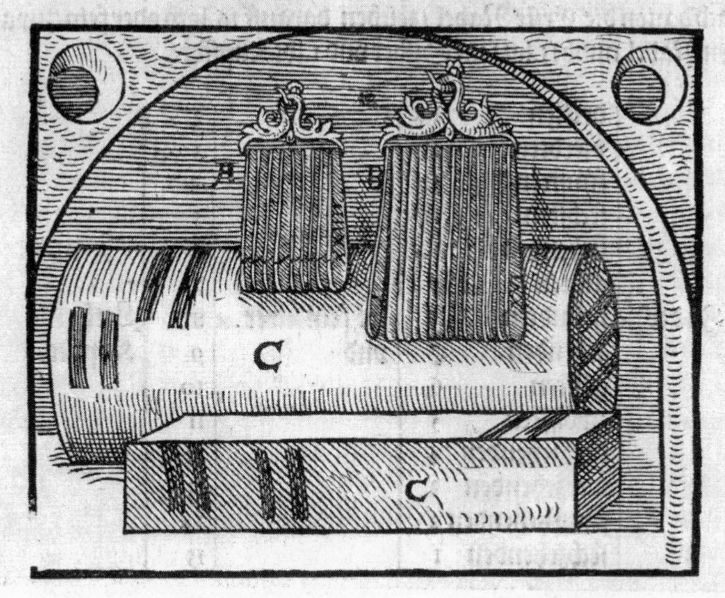

Figure 11. Touch needles and touchstones, from Lazarus Ercker, *Beschreibung allerfürnemisten mineralischen Ertzt unnd Berckwercksarten* (Frankfurt am Main: Georg Schwartz, 1580). (From the Roy G. Neville Historical Chemical Library, a collection in the Othmer Library, Chemical Heritage Foundation. Photo by Douglas A. Lockard.)

out there. When Friedrich died in 1608, two of his personal physicians and a director of his alchemical laboratories took an inventory of "all of the transmutational and medicinal projects, in addition to raw and prepared materials, instruments, furnaces, glasses, crucibles, cupels, and other tools required for work, which are found not only in the [alchemical laboratory in the] Alte Lusthaus, but also in the other laboratories in the palace garden."[29] Because this remarkably thorough inventory listed not only the various processes that were in progress in Friedrich's laboratories, but also the employment agreements and salaries of those who worked there, it affords a snapshot of early modern alchemical laboratories and all of their activities. Twenty-one people were working in Friedrich's laboratories in 1608. This group consisted entirely of men from German-speaking lands (though not all were local to Stuttgart), a few of whom were also identified as pastors, doctors, or barbers.[30] Almost all of the employees were receiving a set salary at the time of Friedrich's death and had clearly defined alchemical tasks, although some details eluded those who compiled the report. A few individuals had arrangements that gave them more autonomy in their work, though this was hardly the norm; these alchemists also had varying levels of responsibility for providing recipes and results. These bustling laboratories, host to a whole range of alchemical projects, formed a vibrant alchemical research center cum manufactory in Stuttgart.[31]

The laboratory in Duke Friedrich's Alte Lusthaus was carefully organized. Friedrich set out the general principles according to which the laboratory was to operate shortly after he set it up for Honauer in 1596.[32] The *Laboranten* were given ducal protection (*Bürgschaft*) in exchange for taking an oath "that they will abide [*haussen*] faithfully and sincerely, as is worthy of faithful servants."[33] Friedrich also stipulated that the palace kitchen deliver soup to his alchemical workers each morning and that the butcher send a daily total of four pounds of meat. In addition, the court provided twenty-five pounds of salted butter every three months, from which each *Laborant* was to receive a slice three times per week in the morning; each worker was also to receive a half goblet of vermouth daily. Friedrich was as scrupulous in setting out rules for managing the laboratory's finances as he was in creating the menu. All deliveries of metals or minerals were to be recorded in writing, as were all payments to the *Laboranten*. The money was kept in a trunk in the basement of the building, and cash was to be given out only in the main hall. There were two keys, each kept by a different person "so that no one can get in without the other."[34] Friedrich was concerned not only that his alchemical laborers be well organized, but also that the finances of this enterprise be carefully managed.

Although each worker apparently made an individual arrangement with Duke Friedrich, differing levels of pay and autonomy made it clear that there was a certain hierarchy within the larger ducal alchemical project. The key divide was between *Laboranten*, who typically worked on recipes that Friedrich assigned to them, and alchemists, who ordinarily provided the recipes or processes themselves. Alchemists, in other words, were expected to have the knowledge to come up with new processes or methods, whereas the *Laboranten* simply executed them. Thus, the inventory tells us, the *Laborant* Daniel Keller "only worked on that which he was instructed to do at the order of His Grace," whereas alchemists like Hans Heinrich Nüschler arrived at court with processes and promises of their own (in Nüschler's case, a process to extract four loth of gold out of one mark of silver).[35] Working as a *Laborant* afforded none of the potential glory or financial rewards of the alchemist. To the twenty thousand gulden Nüschler hoped to receive upon successful completion of his process, Keller earned only three gulden per week, plus two new outfits (*Hoffkleider*) yearly. The *Laboranten* were in a far less risky position, however, because they were not responsible for the success of the process they worked on, financially or intellectually. All Keller had to do was work in good faith on the tincture Duke Friedrich had assigned to him and he would receive his pay, whereas Nüschler, as we have seen, fell into debt when his process failed, and then Friedrich convicted him of fraud and executed him in 1598. One ducal *Laborant* named Alexander Stocker, in fact, exploited the distinction between alchemist and *Laborant* when he was accused of *Betrug* in 1603. He argued that he could not be held responsible for the alchemical promises he was accused of falsely making because he was just employed as a *Laborant,* not an alchemist and thus was not responsible for the success of the processes he worked on. His evidence for his status as *Laborant* and not alchemist was the fact that he had never signed a contract with the duke.[36]

The distinction between alchemist and *Laborant,* then, involved varying levels of pay, risk, and autonomy. At the Württemberg court in Stuttgart, a certain degree of spacial differentiation reinforced the difference between alchemist and *Laborant.* The main laboratory in the Alte Lusthaus seems to have employed mainly *Laboranten* working together under the direction of Duke Friedrich and his managers, but Friedrich supported other laboratories too. With the exception of Honauer, the alchemists who brought their own processes to Friedrich's court did not work alongside the *Laboranten* in the Alte Lusthaus, but rather in their own laboratories (supported by their own *Laboranten*), either in the Neue Spital (New Hospital) nearby or on Friedrich's property (acquired in 1600) in Kirchheim unter Teck. Thus, at various points during Friedrich's reign, the Neue Spital hosted the "great Jewish artisan" Abraham Calorne, who was hired at the Stuttgart court to demonstrate a new process for producing saltpeter, as well as alchemists Petrus Montanus, Andreas Reiche, and Johann Hofrichter.[37] The better-known alchemists Hans Heinrich Nüschler, Hans Heinrich Mühlenfels, and Michael Sendivogius carried out their work outside of Stuttgart in Kirchheim unter Teck. Friedrich's alchemical enterprises, then, were remarkably diverse. He seems not only to have run his own laboratory, but also to have supported several smaller laboratories under the direction of other alchemists.

The divide between *Laborant* and alchemist, however, was not absolute, for Duke Friedrich occasionally allowed *Laboranten* to work on their own projects. Pantaleon Keller, who worked for Friedrich from 1595 to 1608, was hired "to work diligently on that which is . . . given to him, and also to work for himself as industriously as possible on the tincture."[38] The provision that Keller be allowed to work on his own projects as well as those Friedrich assigned to him allowed this *Laborant*/alchemist a certain amount of autonomy, although it came with a catch. In his arrangement, Keller essentially signed away his rights to his own intellectual property. "Whatever good things he discovers," the arrangement specified, "may he communicate them to His Grace."[39] Keller could do his own work in Duke Friederich's laboratory, in other words, and presumably not risk anything if he failed, but the duke retained a right to whatever products resulted. Keller was not alone in making this kind of deal with Duke Friedrich. Hans Jacob Gohel, for instance, admitted that "he worked at his own expense and if he does not manage to do anything useful, he can be dismissed, but in the case that he discovers something useful in his arts, then he is humbly to communicate it to His Grace."[40] Although they could not expect the huge sums that the duke promised to people like Nüschler, practitioners like Keller who occupied a middle ground between alchemist and *Laborant* could more cautiously do their own alchemical research without making promises that might prove impossible to keep.

Beyond the *Laboranten* and alchemists, a number of others came and went in Friedrich's laboratories as support staff and administrators. The duke employed two scribes and bookkeepers to keep track of the payments to *Laboranten* and deliveries to the Alte Lusthaus.[41] The laboratory also required the services of several tinkers and their assistants, as well as of two potters, presumably to produce various crucibles and ceramic vessels and to repair or build furnaces.[42] Occasionally individuals otherwise employed elsewhere at court were sent to work in the laboratory, as was "Andreas the Chamberboy."[43] These laboratory servants too could climb the laboratory ladder, as did Johannes Wagner, who began simply as a servant but was promoted after Friedrich's death to *Laborant*.[44] On the other end of the spectrum of auxiliary employees, Friedrich appointed managers or inspectors to oversee all of the work in his laboratories. It was not unusual for sovereigns (particularly those with larger alchemical projects) to hire someone for this purpose. Vilém Rožmberk, for instance, hired the former mint master Petr Hlavsa to manage his Prague laboratories as well as a woman named Salome Scheinpflugerin to run his laboratory in Třeboň in southern Bohemia.[45] When Friedrich hired his first *Inspektor*, Florian Kappler, in 1595, the two men set out Kappler's duties in a contract.[46] In addition to promising to "inspect diligently and have my watching eye on the *Laboranten*" and to execute any duties the duke might assign to him, Kappler vowed to keep an inventory of all of the materials in Friedrich's Alte Lusthaus laboratory and the uses to which they were being put. In exchange for such services, Kappler was to receive a decent annual salary: one hundred gulden, "fruits," 1,063 liters of rye, 4,253 liters of spelt, 709 liters of oats, 1,175 liters of wine, two outfits, and "candles like other officials."[47] Although his position was obviously essential to the functioning of the laboratory, Kappler was ultimately more an administrator than an alchemical practitioner in his own right.

As these examples from Stuttgart suggest, alchemical laboratories were populated by individuals doing everything from carting around wood to developing and executing alchemical processes. We can make two rough distinctions among these workers (though of course such lines of demarcation were often rather fuzzy in practice): first, between those employed primarily to carry out alchemical work and those hired as support staff, and second, between alchemists and *Laboranten*. The Jesuit polymath Athanasius Kircher articulated this latter distinction seventy years after Friedrich's death in much more concrete terms. As the director of experiments at the Collegio Romano in Rome, Kircher explained the difference between his role and that of his *Laboranten:* "It was my prerogative to order and prescribe their method; theirs to execute my orders." He went on to explain that it would be improper for him to carry out his own experiments, "just as it is unbecoming for physicians to make compositions of medicaments, to open a vein, or insert a syringe, since this is the proper role of the pharmacist or surgeon; just as an architect does not prepare cement or polish marble, but entrusts such things to be done by others."[48] Physicians, of course, *were* increasingly opening up veins in the seventeenth century, just as alchemists of all stripes had long gotten their hands dirty in the laboratory, but that did not make all alchemists equal. Just as artisanal workshops still had both master artisans and apprentices, laboratories often had both alchemists and *Laboranten,* designations which carried different expectations, social status, and rewards. Alchemical laboratories, in other words, were as differentiated as any other kinds of workspaces in the early modern period, in terms of both hierarchy and specialization.

Divisions of Space and Skill

What, then, defined Friedrich's laboratory in the Alte Lusthaus as a particularly *alchemical* space? From the outside it did not stand out especially from the other buildings in the Stuttgart ducal gardens. Even from the inside, it resembled an artisanal workshop as much as anything else. It was well stocked with items that might have been found in any metallurgical or apothecary workshop: "at least three hundred" glass containers, bottles, "sugar glasses" (*Zuckergläser*) for storing solid materials, metal bowls and pots, metal tongs for lifting vessels in and out of the fire, several long ladels for pouring metals, mortars and pestles, scissors, pans, twenty-three "white glazed apothecary boxes," lamps, and even "a beautiful mirror."[49] Such architectural features and everyday items raise an important question: was there anything distinctive at all about alchemical laboratories as compared with other early modern sites of work or the investigation of nature?

Like the Alte Lusthaus, few early modern alchemical laboratories were specifically designed as such; more typically they were installed in whatever space was practical and available to accommodate alchemical work.[50] Laboratories appeared, for example, in a former Augustinian cloister (in Třeboň, southern Bohemia), a former apothecary (in Wolfenbüttel), a palace (in Złoty Stok/Reichenstein, Silesia), a hospital (the Neue Spital in Stuttgart), and a chapel (in Oberstockstall, Austria). Some practicing alchemists even worked in homes, among them Daniel Prandtner and Claudius Syrrus Romanus, both employed by Vilém Rožmberk in Prague.[51] Certainly the majority of alchemical laboratories have eluded documentation, particularly those in ordinary spaces belonging to practitioners who did not record their work. As Steven Shapin has observed, when laboratories were installed in buildings formerly (or simultaneously) used for other purposes, they could bear the conventions and social rules of their original functions as much as they did the associations with their role as laboratories.[52] Thus, a laboratory in an alchemist's own home was likely to be governed by the rules of domestic space and hospitality,[53] while a laboratory in a cloister would carry connotations of sacred space and the conventions and rhythms of monastic life. Given the import of these lingering associations, laboratories might seem to be merely the sum of their parts, rather than clearly defined spaces with special rules and conventions.

And yet by the late sixteenth century certain features had emerged to define alchemical laboratories as distinct spaces in the landscape of late Renaissance work and learning. Some of these features were architectural. Although it is easy to lose sight of them, certain practical considerations had to be taken into account in establishing an alchemical laboratory. Any workspace had to have a hearth or ventilation for multiple furnaces. The need to feed these fires made access to fuel a necessity; this was an important factor in Tycho Brahe's decision to place his laboratory in the basement of his observatory, where fuel would have been easier to deliver.[54] The proximity of Count Wolfgang II von Hohenlohe's laboratory in Weikersheim to a river was probably the result of a practical decision as well, since the laboratory's wastewater emptied out into the moat via channels.[55] Such pragmatic considerations were important and ensured that not just any building was suitable for alchemical work. Architectural features were less important for marking alchemical spaces, however, than were the things inside. A number of objects distinguished Friedrich's Alte Lusthaus as an *alchemical* laboratory: the combination of equipment for both distillation and assaying, the multiple furnaces for various operations, and the presence of materials like antimony and arsenic closely associated with alchemical processes. Although most of the items in Stuttgart, taken individually, might have appeared in an assaying house, apothecary's shop, or smith's workshop, together they would have tipped off any visitor that Duke Friedrich was interested specifically in some kind of alchemy.

On a deeper level, the most distinctive feature of alchemical laboratories was the fact that they occupied a middle ground among artisanal or industrial workspaces and the experimental laboratories that would become closely associated with the "new science" in the late seventeenth century.[56] On the one hand, alchemical laboratories had deep roots in artisanal culture; "arts of the fire" such as distilling, metalworking, and assaying, in particular, all shared practices and materials that linked them to alchemy. Like these related activities, alchemical laboratory work was also often geared toward the production of goods such as medicines, gold, or gemstones that had clear value to princely patrons.[57] In the central European context, moreover, the foundries and assaying houses associated with the mining industry provided a second important framework for alchemical labor. As I have argued, some alchemists directed their work toward innovation and development in central European mining ventures, in which many princely patrons of alchemy were heavily invested.[58] All of these practices, whether artisanal or entrepreneurial, involved not only bodily labor, as Pamela Smith has pointed out, but also a commitment to productivity, even efficiency and profitability.[59] In this sense, the alchemical laboratory closely resembled many of the other workshops of early modern Europe, ranging from foundries to a goldsmith's, jeweler's, or apothecary's shop.[60]

On the other hand, the fact that alchemy was also situated in a long theoretical tradition meant that it was never solely an artisanal practice.

Although alchemical practitioners certainly varied in the level of their sophistication, most brought some sort of theoretical framework to their practical work that would have enabled them to explain, for example, why it was possible to transmute copper into gold or why potable gold was such a potent medicine. In some cases, alchemical practice not only drew on a philosophical understanding of nature, but also contributed to it, creating the kind of feedback between theory and practice that connected laboratory work to natural philosophy. Although, for the most part, Duke Friedrich's *Laboranten* were hardly scholars, their tests of recipes and processes suggest a sort of research program based on experience, if not exactly experiment. Some alchemists did in fact view the operations of the laboratory as tools for the experimental investigation of nature, even as early as the Middle Ages, as William Newman and Lawrence Principe each have argued. Noting that alchemy was "above all an attempt to understand and manipulate matter," Newman has drawn attention to the way in which medieval alchemical texts focused on the use of quantitative measurement (in their use of the precision balance, for example) as "experimental tools for revealing the nature of matter."[61] Principe has explored the experimental dimension of alchemy in terms of specialized apparatus and reproducibility, arguing convincingly that "some alchemists, including the chysopoeians, were highly solicitous in their manual work, carefully employing or designing specialized apparatus for specific goals. They followed the records (though secretive and allusive) of their predecessors and attempted to reduce them to experimental protocols."[62] For those wishing to trace the history of experimental science, the alchemical laboratory had unique features that set it apart from related activities such as distilling or smelting.

Not all alchemists viewed their work as experimental, however, and those who did may have been extraordinary. In general, we might imagine a spectrum of types of laboratories in early modern Europe, encompassing the entire range of artisanal, industrial, and philosophical activities associated with alchemy. The same diversity that characterized alchemy on the macro level could also exist within a single laboratory, however, as the range of practitioners working for Duke Friedrich in Stuttgart suggests. In such cases, space became a crucial means of organizing and differentiating different kinds of alchemical activities. Although the sources are limited and any conclusions at this stage must be preliminary, the evidence from places like Stuttgart is suggestive. While large alchemical laboratories like the Alte Lusthaus were very closely linked to multiple activities, from assaying, distilling, and smelting to transmutation, the articulation of space within such laboratories suggests that practitioners still kept these facets of alchemical practice somewhat distinct from one another.

The Saxon electors' laboratory, located behind the electoral palace in Dresden, exemplifies this point. Because of the strong links between alchemy and mining in Saxony, this laboratory—known as the Goldhaus—served multiple functions as a center of smelting and assaying as well as alchemy. Signaling the importance he attributed to mining, Elector August founded the Goldhaus in 1556, three years after taking the Saxon throne. He continued to introduce new techniques there throughout his reign, probably using it as an experimental laboratory for innovation in the Saxon metallurgical industries.[63] Like the Alte Lusthaus in Stuttgart, the Dresden Goldhaus contained several different rooms that helped differentiate space. The details of a 1598 inventory suggest that each of these spaces housed different kinds of operations. The initial ore-processing operations were clearly carried out in the first and largest chamber, denoted as the "large hall or *Laboratorium.*" This room contained numerous retorts, crucibles, and assaying cupels, in addition to nine smelter furnaces and one blast furnace, both of which types of furnace were used to purify ore. There were also two Lazy Heinzes and one distillation furnace for the more refined processes. A small, locked wooden chest stored nearly thirty books; the majority of them had to do with processing salt, suggesting that the elector's experiments with salt production found a home in the Goldhaus as well.[64]

Beyond the large hall or laboratory were two additional rooms. Adjacent to the large laboratory, a "small round corner room" was filled with the scales, weights, and needles of the sort used in assaying. A small desk covered in "beautiful black velvet" stood in the middle of the room, covered with a number of silver, tin, and leather boxes (with unidentified contents), a few brass dishes, and the material from two assays. The compartments of the desk were filled with silver weights, bits of assayed gold, and balances of the sort Lazar Ercker described in his treatise on assaying. (See fig. 10.) The other furniture in the room contained similar items, suggesting that this room was devoted to assaying.[65] A third vaulted room in the Goldhaus contained instruments for distilling. The officials taking the inventory found eight bowls of vitriol ("with foam still in them"), three glass flasks, a cupel filled with "green water," and another flask filled with foam on a table. Under the table lay two iron jugs, two basins and sieves full of vitriol, three iron rings (for holding up vessels in baths), and tartar, among other things. The rest of the room contained various distilling equipment, including flasks, alembics, a pelican, bottles, and glasses.[66] Many of these vessels contained distilled materials when the inventory was taken, such as "a liquid copper in a crucible," distilled honey, fermented vinegar, and *Scheidewasser*, or parting acid.[67] Taken together, these items suggest that this third room was the primary locus for distilling operations in the Dresden laboratory.

The items listed in the two additional rooms at the back of the Goldhaus suggest that this was where the most sophisticated alchemical work took place. There were several furnaces, including a Lazy Heinz, in the first of these two rooms, or the "large *Laboratorium*" located all the way in the back of the building (not to be confused with the large hall or *Laboratorium* in the front of the Goldhaus). The metals and chemicals in this second laboratory were clearly more refined. Whereas the materials in the front rooms were merely listed in relatively unprocessed forms (e.g., "pellets of copper, may contain silver"), the back room contained materials in more advanced stages, including *sublimatoria*, calcined sulfur, and oil of vitriol. More important, this back room housed a variety of the kinds of powders to which practitioners commonly ascribed transmutory or medicinal powers. On a ledge, for instance, lay "three old single cucurbits, on one of which [was] a label: 'The last *tincture*, from gold.'"[68] The officials taking the inventory also came across "two boxes with a black powder" and "a glass box full of a red powder, with a label [that read]: *Tingeing* powder, which was prepared to *tinge* half gold and silver."[69] A fifth and final, small room, apparently the most hidden of all, was located within this second laboratory. This room also contained written materials. An ivory box containing a "long parchment sheet" sat on the windowsill, while "a booklet in 16° [sextodecimo] in green parchment, written about smelting gold," sat in a tarnished cabinet nearby.[70] These written materials, together with a pair of eyeglasses in another cabinet and an atlas, indicate that the alchemists in this hidden room clearly read, while the "lovely pen with several little boxes and several [pieces of] writing paper" suggest that they wrote as well.[71]

The distribution of objects in each of the rooms in the Dresden Goldhaus suggests that space served to organize and differentiate among facets of alchemical work. Much as in Libavius's imagined Chemical House, which contained an assaying room, a coagulatorium, and several rooms for storage of materials and equipment in addition to the laboratory proper, or "sanctuary," different rooms of the Goldhaus housed separate activities.[72] While practitioners carried out the initial smelting and refining in the large front laboratory, they probably used the other rooms for distillation, assaying, reading, writing, and the more refined alchemical processes. This partitioning of space suggests that practitioners viewed these activities as somewhat distinct from one another, if only because of the equipment they required.

The separation of distilling and assaying from the rest of the alchemical work in Dresden was mirrored in Count Wolfgang II's palace laboratory in Weikersheim. A practitioner of alchemy in his own right, Wolfgang had been using a provisional laboratory since 1592, shortly after moving to Weikersheim. When he began to renovate his residence at Schloß Weikersheim in 1598, transforming a medieval castle into a palace fit for a Renaissance prince, Wolfgang decided to build a more permanent laboratory. For advice in the matter, he turned to his fellow prince and alchemical correspondent, Duke Friedrich in Stuttgart, who only a few years before had completed his own transformation of the Alte Lusthaus into a laboratory. Friedrich, in return, sent to Weikersheim one of his *Laboranten*, Erasmus Mittelbacher, who advised Wolfgang's employee Jörg Spring in devising the plans for the new laboratory.[73]

The Weikershiem laboratory was much smaller than the laboratories in Stuttgart and Dresden, perhaps because it was intended for personal rather than large-scale entpreneurial use. Like the Alte Lusthaus in Stuttgart, the Weikersheim laboratory had two stories connected with an octagonal stairwell. Lacking a hearth, the ground floor was most likely used only for the storage of minerals and other equipment, while the two rooms on the upper floor probably housed the alchemical work. Although we do know more or less what kind of equipment the laboratory contained, unfortunately it is unclear exactly where that equipment was located in each of these upper rooms. Because the large masonry furnace and fixed bellows would have taken up most of one room, Jost Weyer has speculated that the other furnaces (possibly a Lazy Heinz, assaying, and distillation furnaces as well as a "philosophical furnace" for finishing the philosophers' stone) must have been located in the other. Wolfgang's laboratory may have resembled Libavius's imagined Chemical House in this regard, though the simple constraints of space may also have dictated which furnaces went where.[74]

However Wolfgang may have organized the space within his small laboratory, it is clear that he removed distillation work and saltpeter production to separate spaces entirely. Between 1588 and 1608, Wolfgang set up four separate distillation houses on his palace grounds in Weikersheim. These stills were probably used in part for Wolfgang's alchemical work; the proximity of one distilling room to the alchemical laboratory suggests that it, in particular, produced acids and alcohol used in alchemical processes. As Weyer has argued, however, Wolfgang's stills also served another purpose; his employment of a full-time *destillator,* among other things, suggests that his court produced and sold brandy produced from these stills.[75] The court also devoted its resources to the production of saltpeter, which, like distilled alcohol, was useful both for producing acids used in alchemical processes and in nonalchemical goods like gunpowder. As part of a growing saltpeter industry in Europe, Wolfgang's saltpeter works were dedicated primarily to gunpowder production but certainly used for more alchemical goods as well. Because saltpeter works required a good water source, Wolfgang set up his outside the walls of the city near a branch of the nearby Tauber River.[76]

Obviously, practical issues such as limited space and the need for water in part dictated the separation of the stills and saltpeter works from the alchemical laboratory in Weikersheim. More broadly, however, the location of activities like distilling, assaying, and saltpeter production in Dresden and Weikersheim suggests that they may have occupied a slightly different conceptual space as well. These processes were absolutely essential to alchemy, and yet they were not exclusively alchemical, often existing as separate industries in their own right. At both of these courts, they were removed from the most prized (and secretive) elements of the alchemists' work—perhaps even the production of the philosophers' stone—to the more public and accessible areas of the princely palace (even outside the city walls, in the case of the saltpeter works in Weikersheim). This placement suggests that, however essential these processes were to alchemical work, to some extent early modern alchemists and their patrons also saw them as distinct from it.

I have argued thus far for a relatively loose definition of alchemy that can encompass all of its medical, metallurgical, transmutational, and philosophical facets. The architecture of these three alchemical laboratories, however, offers one way to refine this definition. Clearly alchemy included a wide range of activities, sometimes even within a single laboratory. The articulation of space within working laboratories, however, suggests that practitioners and patrons did not think of these activities as one big jumble, but rather as interlocking but discreet pieces of laboratory operations. Space offered one way to think through and separate the different facets of alchemical practice.

Managing Secrecy

Count Wolfgang's removal of distilling and saltpeter production to the more public parts of his palace also raises the issue of secrecy. There is no question that alchemical authors advocated concealment, often developing elaborate practices to protect their art from the uninitiated.[77] Libavius's imagined Chemical House, which relegated the most important alchemical processes to the innermost rooms of the house, offered one model of how space could control the flow of secrets by dispersing them. According to William Newman, "this division of space [in the Chemical House] . . . was not the mere product of casual expedience: It was a conscious adaptation of principles used in alchemical texts themselves to delude the unwary and prevent the overly easy acquisition of knowledge."[78] While these textual principles were easy to implement in an imaginary building, they were more difficult to put into practice in real laboratories because the many different parties involved in entrepreneurial alchemy often had different agendas. While alchemists sought to balance their need for benefactors' support with the traditional injunction to secrecy, princes struggled to control the flow of alchemical arcana within laboratories, courts, and territorial states. Placing the most delicate alchemical operations in the most cloistered rooms was only one of the ways in which princes and practitioners used space to manage secrecy.

Secrecy was an area of compromise, for it had more than one inflection for alchemists and their patrons.[79] It could mean autonomy, for example; in an ideal world, most alchemists probably would have preferred to work alone in their own homes or private laboratories, free from the strings attached to princely patronage. Secrecy could also mean privacy, the desire to work apart from the disruptions of fellow alchemists. Working in close quarters with others could easily breed discontent. Two alchemists working for Vilém Rožmberk on his southern Bohemian estate in Krumlov (or Krumau), for instance, evidently did not get along at all. Jakob Faber and Theophilactus (or Töpfer von dem Trauben) were trying to make the golden medicine *aurum potabile* in 1585 when Faber expressed his concern that someone had been spreading rumors about him. "I have heard from trustworthy people that you received word that I am supposed to have danced . . . and jumped around, day and night, and committed all sorts of treachery," Faber wrote to his patron Rožmberk. The alchemist feared that Rožmberk had, as a result, "turned an ill-humored eye" toward him and offered an elaborate explanation for his behavior. Faber probably suspected Theophilactus as the source of the rumors; he denounced his fellow alchemist in return, lamenting that Theophilactus "destroys so many instruments at the expense of long [hours of] work."[80] Surely such comments are only a glimpse of the much more numerous day-to-day conflicts over fueling furnaces or knocked-over equipment. A desire to avoid these situations by owning a private laboratory, just as much as a principled desire to keep alchemical secrets, certainly lay behind the common refrain in alchemical texts that practitioners should be as well endowed with monetary resources as with learning and experience.[81]

All but the wealthiest practitioners, however, were forced to involve others in their pursuit of alchemical arcana out of the need for funding and access to equipment and workspaces. At best, this kind of compromise meant divulging secrets in exchange for support from a benefactor but retaining a certain amount of autonomy and privacy. Access to a private laboratory was a luxury and a privilege granted only to favored alchemists like Heinrich Nüschler and Heinrich Mühlenfels, who each received space near Stuttgart in the castle at Kirchheim unter Teck. Practitioners at courts where alchemists were less numerous also had more privacy by default. Philipp Sömmering and Anna Zieglerin in Wolfenbüttel were fortunate to find themselves in this position, as were Daniel Prandtner von Prandt and Claudius Syrrus Romanus, two alchemists whose support from Vilém Rožmberk allowed them to work alone in their homes in Prague. At worst alchemists found themselves working alongside numerous other practitioners in large laboratories, as did lesser-known *Laboranten* in the Stuttgart Alte Lusthaus, Dresden Goldhaus, and Vilém Rožmberk's Třeboň, Krumlov, and Reichenstein laboratories. In such cases, privacy, autonomy, and secrecy were difficult to maintain.

Princely patrons involved in entrepreneurial alchemical projects tended to understand alchemical secrecy from a different perspective; they saw it primarily as a question of restricting access to state secrets, rather than the protection of esoteric wisdom. Princes preferred to retain as much control over alchemical secrets (and profits) as possible and took measures to prevent them from falling into the wrong hands. Vilém Rožmberk's nervousness about letting Claudius Syrrus Romanus work alone, for example, manifested itself in a clause in Syrrus's 1577 contract stipulating that the alchemist not allow anyone but Rožmberk into his home.[82] A few years later, a Württemberger barber named Peter Hottenstein and his fellow alchemists encountered similar princely fears when they were caught practicing alchemy outside Württemberg and Duke Friedrich's control. Friedrich's response to Hottenstein and his coconspirators' treachery reveals his concern that someone else might reap the rewards of these alchemists' labor. He required Hottenstein (the only member of the group who was captured) to take an oath swearing that in the future he would inform the duke of his intent to practice alchemy and "wait for His Grace's gracious response" before embarking on any new projects. The ducal *Cammersecretarius* Johann Sattler noted that the duke had instructed Hottenstein to desist from any further alchemical operations, "especially abroad," under threat of corporal punishment. Sattler added firmly that if the alchemist "has an art [to offer], however, may he choose to notify His Grace, [who] will give him plenty to work on."[83]

From the princes' perspective, centralizing alchemical work within one or more laboratories was an effective way to keep secrets under princely control. Even within laboratories, princes like Friedrich did their best to ensure that all secrets flowed through them. The contract for Duke Friedrich's *Inspektor* in the Alte Lusthaus, Florian Kappler, for example, made it clear that, although he was essentially in charge of the laboratory, Kappler was not necessarily privy to everything that went on there. In his signed statement, he recognized this fact, affirming that "Your Grace's *Laboranten* are not obliged to show such things to me or to divulge them when His Grace Himself gives them a process to work on or sends it [to them] in secret."[84] At the same time, large laboratories like the Alte Lusthaus posed their own set of problems. How secret could the Alte Lusthaus really be, for instance, when servants from the palace arrived regularly to deliver soup and fuel, and courtiers strolled in the gardens below? However hermetically sealed alchemical patrons and practitioners imagined their laboratories to be, their location in the semipublic arena of princely courts meant that boundaries were always permeable. In this regard, alchemical laboratories resembled the domestic spaces in which naturalists and philosophers frequently pursued nature's secrets in the early modern period. As a number of scholars have shown, early modern natural philosophers whose primary site of practice was the household always had to balance their scholarly activities with domestic responsibilities and disruptions.[85]

Nevertheless, princes and practitioners alike found several ways in which they could use space to preserve a certain amount of secrecy—in all of its meanings—even within the larger laboratories. The separation of processes in Weikersheim may have been one response to precisely this problem. William Newman has argued that Libavius's Chemical House separated the various processes leading to the philosophers' stone into different rooms for exactly this reason, thereby "insur[ing] that an outside observer could not easily follow the chain of processes leading from furnace to furnace."[86] Newman suggests that this spatial organization may have been intended as a physical instantiation of the alchemical literary technique of dispersion, whereby the various parts of alchemical arcana were distributed throughout a text. The idea was that only an initiate would know how to piece them together; unworthy readers would simply be confused.[87] It may be that the Dresden Goldhaus was organized in this way as well, to prevent outsiders, or even *Laboranten* and servants, from viewing entire alchemical processes at once. It is striking that the more sophisticated processes were housed all the way in the back of the building, tucked away behind the stills and scales in the front rooms. Might these two back rooms have served, like Libavius's "sanctuary," as "a secret recess away from the other parts of the workshop" in which to perfect the philosophers' stone away from the eyes of observers?[88] Unfortunately, hints are scarce about how information flowed through alchemical workspaces; it is unclear, for instance, whether the back rooms of the Goldhaus were off-limits to the smelters and distillers in the front. Nonetheless, architectural arrangements for alchemical work suggest how practitioners and patrons managed the inevitable compromise between the desire for privacy and secrecy and the reality of practical conditions that made ideals nearly impossible to achieve in practice.

The Danish astronomer Tycho Brahe was exceptional in this regard. With vast personal resources and the patronage of the Danish king, he was able to custom build his own research center on the island of Hven. Tycho was indeed secretive about his alchemy, at least in some respects. Tycho's desire for secrecy was rooted in a belief he shared with Libavius: that certain knowledge—such as how to make the philosophers' stone—should not be too widely publicized. Tycho realized, too, that the judicious release of secrets could have advantages. "I shall not shrink from discussing these matters openly with noblemen and princes," he wrote, "as well as other distinguished and learned men who are interested in such questions and have some knowledge of them; and, at the appropriate time, I shall share some things with them, provided that I am convinced of their goodwill and of the fact that they will keep these things secret." Even so, alchemical secrets were not for everyone: "For it is not expedient or fitting that such things become common knowledge," Tycho continued.[89] Extremely skilled at maneuvering within European patronage networks, Tycho believed that the primary value of alchemical secrets rested in their role as unique gifts. As Jole Shackelford points out, this encouraged Tycho's secrecy, since "the value of such a gift was in its singularity and would have been diminished with publication."[90] By placing his entire scientific enterprise (and not just his alchemical laboratory) on an island, Tycho could control just who had access to his secrets, thus maximizing their value.

As a potentate with vast resources, the support of the Danish king, and an entire island (complete with inhabitants) at his disposal, Tycho Brahe was hardly a typical alchemist. Few others could marshal the means to realize a space for the pursuit of natural knowledge like Uraniborg, which came as close as anything to a custom-built, idealized structure for the study of nature. Images could visually encapsulate rich polemics on the figure of the alchemist, and designs for laboratories such as Libavius's Chemical House could communicate important principles of the alchemist's work and place in society; crucially, however, these imaginary laboratories never had to be built and used in the messy indeterminate world of alchemical laboratory practice. The working laboratories that dotted the Holy Roman Empire were complicated by the fact that they had to balance ideals and realities, as well as the priorities of both patrons and practitioners. They were the product of compromise, coexistence, and appropriation as much as alchemical tradition and principle.

Like the moment in which alchemists and their patrons signed contracts, then, the establishment of a laboratory required all kinds of decisions in order to put alchemy into practice. The organization of space was a central issue, and the choices made reveal a great deal about what was important to alchemists and their patrons. Practical issues like available space, ventilation, and access to a hearth probably had the most influence over what laboratories ended up looking like. As a result, laboratories were located in buildings that had once been all kinds of other things: summer houses, cloisters, apothecaries, and hospitals. Once they were converted, however, such spaces tended to be dedicated alchemical workspaces, housing the entire range of practices associated with that art.

Attention to the use of space in alchemical laboratories also reveals two other priorities, however: the differentiation of alchemical work along lines of both hierarchy and practices, and the multiple valences of secrecy. In Stuttgart, Friedrich frequently gave alchemists (as opposed to *Laboranten*) their own laboratories, reflecting their autonomy and status vis-à-vis more common alchemical workers. This spatial reinforcement of hierarchy underscores the fact that not all alchemists were alike, but rather differentiated along lines of autonomy, responsibility, and skill. These distinctions turned out to be crucial, resulting in very different expectations, risks, and rewards. Likewise, even within large laboratories that housed several different kinds of activities at once, particular types of alchemical operations—such as distillations and the production of transmutory agents—appear to have been kept separate, suggesting that it was important to those involved in alchemy to distinguish among its multiple parts. Second, space provides some insight into secrecy. Just as alchemical texts repeatedly emphasized, secrecy was important to alchemists and their patrons, who designed laboratory structures to preserve it. Multiple keys, restrictions on where *Laboranten* could practice (and for whom), and explicit statements about who could enter workspaces all served to control alchemical secrets. Nevertheless, the desire for secrecy meant different things to alchemists and their patrons. Alchemical practitioners, whether alchemists or *Laboranten,* saw one set of advantages in working alone, even if it rarely turned out to be possible in practice. Beyond adherence to ancient injunctions not to divulge privileged knowledge, privacy meant that their work was less likely to be disturbed by dancing and jumping laboratory partners. It also meant that their secrets remained under their own control, and they were not beholden to the whims of demanding and powerful patrons. Patrons, on the other hand, viewed secrecy as a way to preserve state secrets and prevent them from falling into the wrong hands, corralling practitioners within spaces under centralized princely control whenever possible.

No doubt we still have a great deal to learn about alchemical laboratories in early modern Europe. This chapter has drawn attention to the particular issues surrounding princely alchemical laboratories, but this is only one type of laboratory that must rest alongside studies of many others: chymists' private laboratories, imaginary laboratories in paintings and engravings, and others we do not yet know about. Perhaps future research will uncover a private urban laboratory, where an alchemist's molten product shared the hearth with his wife's soup. Certainly any number of other types of alchemical workspaces are left to discover. This, in the end, is the answer to the question, What did an alchemist's laboratory look like? Beyond practical issues, alchemical laboratories were no doubt as diverse as alchemy and its practitioners in early modern Europe. Each individual laboratory was the result of a unique set of decisions that reflected the constellation of priorities and practical realities that shaped the desire to practice alchemy and devote space and resources to it. Each laboratory, in the end, embodied all the tensions of alchemy itself.

CHAPTER SIX

Betrüger on Trial

In 1586, eleven years after Duke Julius of Braunschweig-Wolfenbüttel executed Philipp Sömmering, the duke received a letter from Count Palatine Richard zu Simmern. Richard succinctly described both the seduction and the frustration of alchemy: "I know well that for a long time Your Dearest has delighted and amused Yourself in certain arts, especially with *Chimia* or *Alcamia.* And this despite the fact that it cost Your Dearest more and more and also Your Dearest was cheated by vagabonds and impostors (*landstrichern und betriegern*) of this alchemical art. . . . Indeed I have also dealt with these things for more than twenty years and [have been] rather put upon by the same people, such that I resolve repeatedly to forget about alchemy, but then once again a new process comes my way, which causes me to work (*laboriren*) again."[1] Such private letters evoke the pervasive fear that the Holy Roman Empire was in fact swarming with *Betrüger* seeking to take advantage of eternally hopeful patrons like Richard zu Simmern.

In the late sixteenth and early seventeenth centuries, these fears were confirmed most dramatically, as princes not only tried and convicted a series of alchemists for the crime of *Betrug* but sometimes also, as in Sömmering's case, executed them in highly publicized and elaborate ceremonies. Although only a minority of alchemical practitioners actually ended their careers on the gallows, their fates made a strong impression on contemporaries and later historians alike. Duke Wilhelm V of Bavaria, Friedrich I of Württemberg, Julius of Braunschweig-Wolfenbüttel, Count Wolfgang II von Hohenlohe, and Elector August I of Saxony together initiated at least eleven official inquiries into alchemical fraud between 1575 and 1606. Not all of these cases resulted in a death sentence. Abraham Calorne and David Pirkheimer, for example, escaped punishment in

Stuttgart by fleeing town before Duke Friedrich brought charges, while the outcomes of his cases against Alexander Stocker and Michael Heinrich Wagenmann remain unclear. In Weikersheim, moreover, Count Wolfgang II mitigated Michael Polhaimer's death sentence, commuting it to imprisonment until the alchemist could pay his debts.[2] Still, German princes staged at least nine executions of alchemists in this period. As we know, Philipp Sömmering and his collaborators died rather spectacularly in Wolfenbüttel in 1575. The Munich executioner wielded his sword (rather ineffectively, as it turned out) to behead Cypriot alchemist Marco Bragadino in 1591.[3] In Stuttgart, Duke Friedrich I used a huge iron gallows in the center of the city to hang Georg Honauer in 1597, Petrus Montanus in 1600, Hans Heinrich Nüschler in 1601, and, finally, Hans Heinrich von Mühlenfels in 1606.[4]

At first glance, such convictions seem fairly straightforward, proof that deliberately deceptive swindlers did try to con credulous princes in the early modern period. The characters from humanist satires at the beginning of the century seemed to have come to life at the end. And yet the trials, convictions, and reactions to them were much more complex, revealing all the subtleties and nuances that had accumulated around the notion of alchemical fraud by the early seventeenth century. As princes struggled to determine whether alchemists had simply failed or had committed a crime more sinister, they revealed the way in which their own conceptualization of alchemy as an entrepreneurial art had created the conditions that brought the literary *Betrüger* to life. By using contracts to transfer as much risk as possible to alchemists, princely patrons, in effect, set them up to bear the full consequences of failure. Alchemists accused of *Betrug,* meanwhile, resolutely insisted on their own narratives about what had gone wrong, even as they faced torture and lengthy interrogations designed to elicit confessions.

These executions did not go unnoticed among alchemical authors, either. Alchemists such as Leonhard Thurneisser, of course, had long defended alchemy in print by differentiating between the "true" alchemists and the false *Betrüger.* By the early seventeenth century, however, the stakes involved in this distinction were even higher. First of all, the execution of real alchemists for *Betrug* drew unprecedented levels of attention to the putative connection between alchemy and fraud. If humanists and artists had introduced the figure of the *Betrüger*-alchemist to a new literate urban audience, then the executions carried the persona to an even wider audience still; executions, after all, were meant to be public in the widest sense during the early modern period.[5] This meant that if anyone had missed the effort to connect alchemists with fraud in print, he or she could see that connection firsthand on the gallows and in the broadsides that commemorated executions. In response, alchemists had to work even harder to differentiate themselves from the despicable *Betrüger* who threatened to throw alchemy into disrepute for once and for all. The obsession with identifying and unmasking fraudulent alchemists, therefore, appeared in alchemical literature with a new urgency in the early seventeenth century in the wake of fraud trials. Books such as the alchemist Michael Maier's *Examen fucorum pseudo-chymicorum detectorum et in gratiam veritatis amantium succincte refutatorum* (Swarm of drones; or, A critical examination of the unmasked and quickly refuted pseudochymists for the sake of the lovers of truth, 1617) attempted to dispatch the "*pseudo-chymici*" by identifying them and exposing their tricks, while one of the early Rosicrucian texts, the *Chymische Hochzeit Christiani Rosencreutz* (*The Chemical Wedding of Christian Rosenkreutz*, 1616), reflected the recent executions directly with a scene in which the false alchemical pilgrims were tried and executed for their crimes.

What then did early modern alchemists, patrons, and observers mean by alchemical fraud in the context of actual fraud trials? This chapter examines *Betrug* from the perspectives of the three main parties involved: the princes who tried and executed "false" alchemists, the learned alchemists who wrote polemical treatises against them, and the so-called "frauds" (*Betrüger*) themselves. The first two groups clearly distinguished between "real" and "false" alchemists, although they each understood this category differently. What of the third group, the *Betrüger*? Is there evidence of their devious intentions? Did they self-consciously set out to defraud their patrons, as their accusers maintained, and did they actually resort to the tricks that appeared in alchemical books? Or is it possible that they, too, believed that what they were doing was indeed legitimate alchemy, even if they couldn't always get it quite right?

Alchemy and the Law

Princely patrons did not punish unsatisfactory alchemical practitioners indiscriminately. Like all citizens of the Holy Roman Empire, sovereigns were bound by laws that carefully regulated the administration of justice. In particular, the early modern alchemical trials were rooted in a tradition of legal writing about alchemy that extended as far back as the thirteenth century. In one important thread of medieval canon law, theological concerns about the art were primary. Alchemy appeared most often as a point of comparison in discussions of the limits of the power of demons. As William Newman has shown, this link between alchemy and demonology centered on the shared notion of changing or transmuting "species": if alchemists could transmute one "species" of metal into another, then so too could demons and witches also change "species," that is, that they could change shape from humans into animals. At heart, this was a question of whether it was legitimate to believe that anyone other than God, whether alchemist or demon, had the power to break the rules of nature.[6] Those who wished to denounce the belief in demonic shape-changing as heretical tended to condemn alchemical transmutation as well, whereas those who wished to unleash the powers of demons were forced also to concede alchemists' claims to transmute metals.[7] In this important thread of late medieval canon law, therefore, the theological fortunes of alchemists, demons, and witches rose and fell together.

Medieval scholars of canon law were not exclusively focused on alchemy's connection to demons, however; some joined Pope John XXII (r. 1316–34) as well as a number of secular rulers in introducing a second thread into the discussion of alchemy's legality. This line of argument centered not on alchemy's problematic theological status, but rather on a much more worldly set of issues: the legality of alchemical gold in the marketplace and its circulation as specie. It seems to have been this set of issues, in fact—rather than the mounting paired concerns about the powers of alchemists and demons—that prompted Pope John XXII to issue a 1317 decretal on the subject of alchemy known as *Spondent quas non exhibent* [They promise that which they do not produce].[8] The decretal begins with a skeptical evalutaion of alchemists' claims: "Though there is not such a thing in nature, they [i.e., alchemists] pretend to make genuine gold and silver by a sophistic transmutation." Interestingly, John did not dwell on this point about false transmutations, which was of such concern for Petrarch and others, but he moved quickly to the specific consequences of alchemical deceit: counterfeit coins minted from falsely transmuted gold and silver.[9] "To such an extent does their damned and damnable temerity go that they stamp upon the base metal the characters of public money for believing eyes," John continued, "and it is only in this way that they deceive the ignorant populace as to the alchemic fire of their furnace." The remainder of the decretal outlines a series of punishments for counterfeiting gold or silver, ranging from fines "to be used for the poor" to branding "with the mark of perpetual infamy." For those who actually minted coins with the false gold or silver, however, the punishment was worse: "Those, however, who in their regrettable folly go as far as not only to pass monies thus made but even despise the precepts of the natural law, overstep the limits of their art and violate the laws by deliberately coining or casting or causing others to coin or cast counterfeit money from alchemical gold or silver, we proclaim as coming under this animadversion, and their goods shall be confiscate, and they shall be considered as criminals. And if the delinquents are clerics, besides the aforesaid penalties they shall be deprived of any benefices they shall hold and shall be declared incapable of holding any further benefices."[10] Pope John XXII did not dwell on the "art-nature question"—whether alchemists' abilities were able to surpass nature—or even whether the belief in the transmutation of metals was tantamount to the heretical belief in demons' powers of shape changing; rather, he sought to stop the circulation of false alchemical gold and silver as coinage. The clear concern of this decretal with counterfeit money is a reminder that, even among popes, secular concerns about alchemy were often as prominent as more theological objections.

Several other canon lawyers and secular sovereigns made this same connection between alchemy and counterfeiting in the fourteenth and fifteenth centuries. The central issue was the sale of false gold and silver as if it were "real" or using it to counterfeit coinage; the practice of alchemy *per se* was not thought to be illicit, except insofar as it could contribute to counterfeiting. Among scholars of canon law, Andrea de Rampinis of Isernia, Baldus of Perugia, Fabianus de Monte S. Severini, and Alberico da Rosciate of Bergamo all took up the issue in the early fourteenth century.[11] Rampinis, for example, argued that it was illicit for alchemists to sell false gold as real gold or to sell it for minting coins without the prince's permission. Following Rampinis, Rosciate concurred that selling false alchemical gold as true gold was illicit; if the alchemical gold was true, however, he argued that its sale was unproblematic. Princes also issued a series of laws in the late Middle Ages designed to stop the production of coins through false transmutation. In 1380, for example, King Charles V the Wise of France (r. 1364–80) banned the practice of alchemy and the possession of alchemical equipment altogether, presumably out of the same fear of counterfeiting that motivated Pope John XXII's decretal. Likewise, in 1403–4 King Henry IV of England issued a law against counterfeiting, making the offence punishable by death and confiscation of property.[12]

Secular rulers in the fourteenth and fifteenth centuries who took up the issue of alchemy's legality, however, walked a fine line between condemning debased coinage and minting it themselves. As Barbara Obrist has noted, rising administrative costs around 1300 led some European princes to begin to debase coinage.[13] When King Philip IV the Fair of France

(r. 1285–1314) began to mint coins with a lesser amount of silver in them, the English responded with legislation in 1299 designed to keep these "false" coins from the realm. By the mid-fourteenth century, the English kings introduced golden coins to counteract debased silver coinage; in turn, counterfeiters hoarded these golden coins, replacing them with coins of lesser value. Facing debased coinage issued by both foreign rulers and local counterfeiters, several English sovereigns apparently decided to get into the game themselves, and they turned to alchemists for assistance. In 1329, therefore, King Edward III of England (r. 1327–77) issued an order to bring two alchemists named John le Rous and William of Dalby to court. "By that art [of alchemy]," Edward proclaimed, "and through the making of metals of this sort, they will be able to do much good for us and for our kingdom."[14] Henry VI (r. 1422–61) and Henry VII (r. 1485–1509) followed suit. After Henry VI's unsuccessful general call to the nobility, clerical orders, physicians, and natural philosophers to help replenish princely coffers, he issued a series of royal privileges to laymen from 1444–60 authorizing the practice of alchemy in the service of the crown. In 1452 he also created a commission to arrest anyone practicing alchemy without such a license, attempting to ensure that the crown was the only producer of alchemical coins.[15] In short, the laws that sovereigns issued against the alchemical production of coins in the late Middle Ages must be read in the broader context of royal fiscal policy, for sovereigns only condemned the use of alchemy to produce coinage when someone else did it. One king's counterfeiter, it seemed, was another king's master of the mint.[16]

Ruling authorities, particularly in the Holy Roman Empire, had other worries beyond the state of their treasuries. In 1493 the city council of Nuremberg passed an ordinance that reflected a concern with the spiritual and social health of the urban commune, as well as the finances of its citizens. Underpinning the ordinance was the conviction that alchemical transmutation actually was possible, but extremely difficult. "Although alchemy was named and regarded as an art by the teachers of Bible," the ordinance read, "and although many wish to learn how to do it and practice it, nonetheless this art is so sophisticated and so hidden that as far back as human memory goes hardly anyone has managed to truly understand it or really practice it." The ordinance went on to assert that most alchemists, in fact, were either naively incompetent or dangerous *Betrüger*, with the result that "many people are betrayed, their fortunes put in jeopardy, and are led away from God by those who go around bragging and claiming that they are men of this art. Furthermore, through the costly experiments [people] are brought into not a little ruin, so much so that nothing can be done about it." Like John XXII's decretal from 1317, the Nuremberg city council opted to counter alchemical malfeasance, whether due to incompetence or deliberate deception, by criminalizing it: "In order to please God Almighty and in light of the damage being inflicted because of the practice of this art, the wise council of this city for the common good and piety orders that: If someone, by the enticement and practice of this art, brings someone else to financial ruin, and if, when he is denounced and brought before the court, these charges are proved to be true, then he shall be penalized in body and goods in accordance with the results of the proceedings and according the judgment of the Council."[17] The Nuremberg city fathers, in other words, were moved to regulate the misuse of alchemy to preserve "the common good and piety" of the urban commune. Like their contemporary Sebastian Brant, they too worried about alchemy's seductive powers, its ability not only to ruin individuals and families financially, but also to lead them "away from God."[18]

From Pope John XXII to the Nuremberg city council, then, late medieval rulers who issued laws and ordinances against alchemical counterfeiting made it clear that not only the church, but also the state had an interest in regulating the practice of alchemy. If theologians and canon lawyers focused on whether it was heretical to believe in alchemists' claims to transmute metals, secular rulers (together with some scholars of canon law) demonstrated that it was not the practice of alchemy *per se* that was at issue, but rather its misuse. In an early attempt to monopolize alchemical knowledge for the good of the state, one set of laws focused on alchemy's purported role in counterfeiting, declaring that if alchemy was to be used in the production of coins, then it must be done only under the auspices of the sovereign. The ordinance from Nuremberg, however, highlighted a different set of potential alchemical misdeeds, those of the *Betrüger* who threatened to ruin the financial and spiritual wellbeing of Nuremberg's citizens, for whom the city council took responsibility. In the early modern period, it was alchemy's position at the nexus of the state and communal interests that determined the fates of alchemists who found themselves on trial in the courts of the Holy Roman Empire.

The Princes' Perspective

Interestingly, ordinances specifically targeting alchemy seem to have disappeared in the early modern period as alchemy was more fully integrated into European society and culture. When Holy Roman Emperor Charles V issued a new penal code for the Holy Roman Empire in 1535, alchemy did not appear as a distinct crime; this meant that when princes sought to prosecute false alchemists, they had to fit alchemical crimes into much more general categories. Tellingly, they chose to prosecute false alchemists not for witchcraft, heresy, or magic, but for the thoroughly worldly crime of fraud (*Betrug*). According to the imperial penal code, known as the *Carolina,* fraud was a broad category that encompassed offences such as counterfeiting coins, falsifying seals or official papers regarding the rights of lords, using false weights or otherwise misrepresenting products for sale, and moving stones marking property lines.[19] Although the *Carolina* prescribed slightly different punishments for each type of fraud, these crimes all revolved around the dangers of misrepresentation in the context of economic exchange. As Sebastian Brant imagined, alchemists too may have committed this type of fraud, selling false tinctures and powders to unsuspecting burghers in the cities of the Holy Roman Empire by passing them off as the true philosophers' stone. Prosecuting alchemical fraud as this type of commercial *Betrug* underscored early modern alchemy's commercial roots and the increasing marketability of alchemical goods.

If in theory the notion of alchemical *Betrug* drew on this broad legal definition of commercial fraud, however, it is important to note that the alchemical fraud trials that actually took place around 1600 focused on a much more limited type of alchemical transaction. The princes of the Holy Roman Empire did not put alchemists on trial for selling false philosophers' stones in the marketplace, nor did they pursue alchemists for circulating counterfeit coinage, the two concerns of the late medieval ordinances dealing with alchemy. Rather, the sixteenth- and seventeenth-century alchemical fraud trials zeroed in on those alchemists who were thought to have defrauded a very particular type of customer: the *princes* of the Holy Roman Empire. Those who went to court for alchemical *Betrug* were invariably the entrepreneurial alchemists, those who entered into contractual relationships with princes but in the end failed to deliver the alchemical goods and processes that they promised.

If the *Carolina* made deliberate deception central to the general crime of *Betrug,* the role of deception was more complicated in alchemical fraud trials. In principle, princes distinguished between incompetence and deception and sought to detect the difference in their pursuit of alchemists who failed to fulfill their contracts. Unlike the members and patrons of other professions, princes did not have recourse to independent experts such as a guild or *protomedicato* to help them evaluate alchemists' work. Instead, they relied on their own judgment and the guidance of their advisers and managers.[20] For central European princes, rumor and denunciation were sometimes enough to raise suspicions about alchemical foul play. The alleged misdeeds of Alexander Stocker and his colleagues came to Duke Friedrich I's attention in June of 1603 when a very public fight between Alexander and his brother, Hans Jacob, turned ugly. Hans Jacob and Alexander caused such a great *Tumult* on a rooftop one morning that twelve townsmen fled the ruckus, but not before Hans Jacob "let it be heard" that Alexander was a flatterer who had admitted that "he wanted to make an ape out of His Grace and was leading him by the nose." What is more, Hans Jacob claimed, Alexander had admitted that, not only had he not even begun his alchemical work, he did not even know how to do it and had seduced Duke Friedrich with a false alchemical process. Further, Hans Jacob clamored, Alexander was a "godless" man who should already have been dead, for his family had saved him from execution in Strasbourg at the very last minute, buying his freedom even as he stood on the ladder with the noose around his neck. Needless to say, when Duke Friedrich received the police report describing these events, he ordered Alexander Stocker and his entire retinue arrested and launched an inquiry into the charges.[21] Whether or not such accusations were true, they could be disastrous for the accused. As in the case of Philipp Sömmering, the mere allegations of fellow courtiers could initiate a series of events with tremendously serious consequences.

A practitioner's flight could also suggest alchemical misdeeds, as the story of the ill-fated Georg Honauer illustrates. As Duke Friedrich told the tale in his warrant for the alchemist's arrest, Honauer had announced himself in the fall of 1596 first in writing and then in person as both an alchemist and as Lord of Brunhoff and Grabschutz in Moravia. Friedrich ordered his assayers to test the alchemical claims, and evidently was pleased with the results of Honauer's transmutations.[22] The large transmutation, however, still lay ahead. The duke originally wanted the alchemist to produce 200,000 ducats worth of gold out of iron, but when Honauer said that he had only enough tincture to transmute thirty thousand ducats worth, the two agreed on an initial production of seven thousand gulden over three months, followed by 36,000 ducats every month indefinitely.[23] In preparation for a final, large-scale demonstration, Friedrich imported to Stuttgart all of the available iron from his armory in Mömppelgart.[24]

In the midst of this large test, however, Honauer evidently panicked. He and his stablemaster simply ran away, leaving behind a huge debt in Stuttgart (somewhere between thirteen and nineteen thousand gulden). After a long and fairly complicated diplomatic wrangle, Honauer was extradited back to Friedrich's Duchy of Württemberg and the interrogation began in March 1597. Honauer's case was exacerbated by the fact that he was convicted not just for alchemical *Betrug*, but also for pretending to be a Moravian lord (the court disputed his claims to nobility as Herr zu Brunnhoff); the social and alchemical deceits, it seemed, went hand in hand. Ever resourceful, Friedrich used the iron he had imported for Honauer's alchemical work to build a thirty-foot high iron gallows on the Schloßplatz in Stuttgart. According to a contemporary woodcut, the gallows weighed over a ton and cost Friedrich three thousand gulden to construct. The gallows itself, the woodcut's printer claimed, was gilded and Honauer wore a golden garment, both serving as a symbolic reminder of his false promises. On 2 April 1597, Honauer was hanged for his crimes. As the text on this woodcut bluntly concluded, "he should learn to make gold better" (fig. 12)[25]

Interestingly, Friedrich's accusation of *Betrug* was independent from his judgment of Honauer's ability to transmute metals. In fact, the Duke continued to believe in Honauer's competence as an alchemist even after he issued the arrest warrant branding him a fraud. In fact, as the duke confessed later to his fellow alchemical enthusiast, Landgrave Moritz of Hessen-Kassel, Friedrich chased after the runaway alchemist so vigorously precisely because he was so sure that Honauer could transmute metals: "The reason we are pursuing the deliberate and utter *Landbetrueger*, the supposed Brunnhoff, so heavily and fervidly is because we know well, plus an assay evidently confirms, that he has the tincture with him . . . and he knows how to do this art [of alchemy]."[26] Duke Friedrich concluded that Honauer was a *Betrüger*, in other words, not because he believed the alchemist to be incompetent, but because he withheld the valuable secret from his patron in violation of their agreement. The fact that Friedrich had converted his Alte Lusthaus into a laboratory for Honauer and emptied the iron from his armory must have infuriated the duke all the more.

The particular constellation of factors in Honauer's case was unusual, recalling the severity of Sömmering's condemnation. More typically, alchemical incompetence and the failure to fulfill a contract went hand in hand; in other words, when alchemists did not fulfill their contracts, it was usually because they simply could not. Incompetence alone did not necessarily have such severe consequences. Michael Heinrich Wagenmann, for instance, wisely confessed to Duke Friedrich his inability to fulfill to his contract with an honesty that probably saved his life. He admitted that his tincture, which he had purchased from a man from Munich, simply did not appear to work, despite his long hours of labor.[27] Unfortunately, because Duke Friedrich's marginal note on Wagenmann's disclosure was

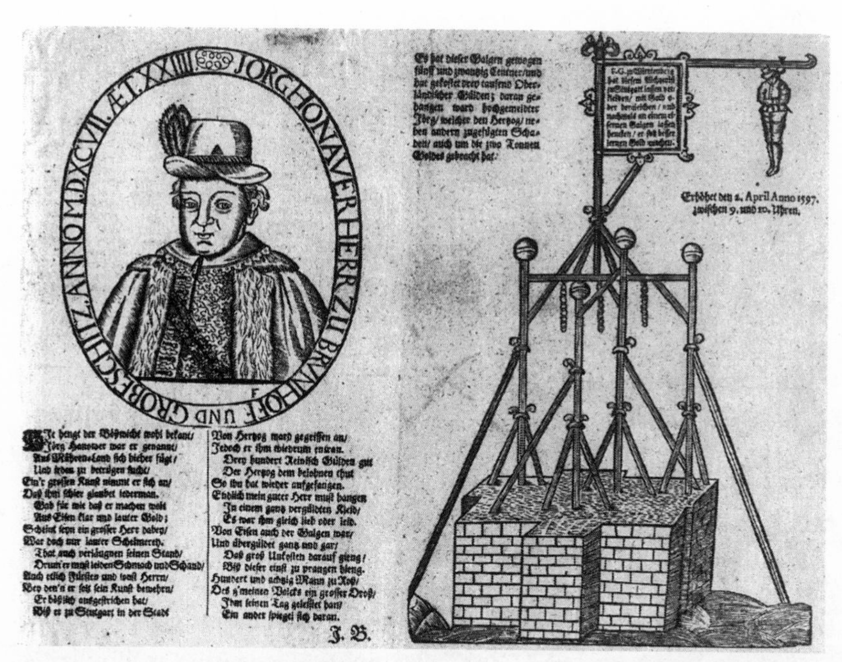

Figure 12. The execution of Georg Honauer, ca. 1597. Verses by Johann Beck (?). (Graphische Sammlung, Germanisches Nationalmuseum, Nürnberg.)

crossed out and is now illegible, the would-be alchemist's fate is unknown. He does not appear to have been executed, however, and might even have been the same "Wagemann" who later worked as a ducal pastry chef.[28] He may also have simply repaid his debt (which his contract stipulated as a penalty)[29] or landed in prison until he could do so. Debtor's prison was also the fate of Andreas Reiche, whom Friedrich employed in 1601 to make the universal tincture and the philosophers' stone. After working for over a year without success, Reiche was imprisoned until he could repay his debts. Friedrich's successor Johann Friedrich and Johann Friedrich's mother Duchess Sibylla eventually released Reiche in 1610 on the condition that he serve as the family's alchemical tutor.[30]

In a few cases, alchemical incompetence ultimately did lead to a desperate act of deception. Such was the case for Hans Heinrich Nüschler, the blind burgher of Zürich who worked for Friedrich from 1598 until 1601. The alchemist first signed a contract with the duke stipulating that Nüschler would receive twenty thousand gulden if his process for transmuting silver into gold were proven to be successful. When, as Nüschler put it, "the end [of the process] did not want to work out," he bought a new tincture from some young men who claimed that it would (rather fantastically) transmute one part lead into seven hundred parts gold. "With joy I moved residences," Nüschler later wrote, "set myself up in Freihoff," where Friedrich had provided him with a laboratory. This good fortune quickly turned, much to Nüschler's horror. "When I tested the powder," he reported ominously, "it was nothing. Then I was in the most serious peril."[31]

Upon discovering that the tincture with which he hoped to fulfill his contract was useless, Nüschler resorted to a much more dangerous tactic. He asked his brother surreptitiously to mix three *quentlein* of gold (approximately eleven grams) into the gold that court officials were to test for quality. Hoping to convince Duke Friedrich in his plea for mercy that he had done this not out of treachery, but out of desperation, Nüschler later wrote: "Out of great fear I contrived against His Grace the scheme with the three *quentlein* of gold, which were to be used with the help of my brother. . . . I dedicated everything that I still had to that cause, but no luck would come. . . . Finally, what I did in the tests was out of fear and terror. . . . I regret my sins from my heart, and have great remorse about it all, and so I beg . . . that I may receive a gracious judgment from my judges."[32] Nüschler's judges ultimately found this argument unconvincing, and convicted the alchemist for *Betrug*. On 19 July 1601, Duke Friedrich used the iron gallows built for Honauer for the third time to hang the Swiss alchemist.

Interestingly, neither Friedrich nor Nüschler chose to pursue the "young men" who sold Nüschler the false tincture, so the alchemist had to take responsibility for it himself.[33] Nüschler had clearly hoped that his explanation of his actions would ameliorate his punishment; a planned, intentional swindle, he assumed, was not the same as resorting to tricks only as a way out of a desperate situation. From the legal perspective, however, the alchemist's state of mind was irrelevant. Quite simply, Nüschler had failed to fulfill his contractual obligations to his patron, and he had employed deceit to cover up his failure. The administrator who handled the case reduced it to its barest elements, recalling it later as essentially a matter of "a contract, which [Nüschler] had arranged with our Gracious Prince and Lord . . . for twenty thousand gulden."[34] From the perspective of the courts (both legal and princely), in other words, the heart of trials for alchemical *Betrug* was the contract: whether or not it had been fulfilled, and whether or not it had been fulfilled honestly and truthfully.

Even if alchemical *Betrug* constituted breach of contract, it still is not obvious why some princes punished alchemists like Honauer and Nüschler so severely. In general, a fraud conviction did not automatically lead to the gallows in the Holy Roman Empire. If the punishments that false alchemists received seem disproportionately severe, it is because they broke contracts with and deceived princes (and thus the state), not just any citizens of the empire. In doing so, they threatened to damage not only the personal alchemical aspirations of their noble investors, but also the political and economic projects on which princes increasingly staked their authority in the sixteenth century. In this sense, alchemical fraud was a crime against the state, positing the prince as victim. Sovereigns from Pope John XXII to King Henry VI of England had long made this claim, of course, but by the early modern period, the contracts that princes and alchemists signed personalized the relationship between alchemist and the state and made it much more concrete. When this relationship went awry, the fact that princes punished alchemists so ruthlessly makes it clear that they had a great deal at stake in their pursuit of entrepreneurial alchemy.

Proper Bees and Rotten Drones

The publicity surrounding the trials and executions of fraudulent alchemists around 1600 opened up the issue of alchemical *Betrug* to a broad audience and invited comment. If princes' reasoning in prosecuting alchemists for fraud was shaped largely by contractual obligations, alchemical authors writing in the years around 1600 articulated a more multifaceted response. Of course alchemical literature had long included discussions of the differences between "true" alchemists and *Betrüger*. In the wake of these trials, however, advocates of alchemy's social and intellectual promise found it more imperative than ever to distinguish themselves from their more disgraceful peers. Thus, at the beginning of the seventeenth century authors took up the issue of the true and false alchemist with a new urgency that focused on the connections among moral depravity, entrepreneurial alchemy, and *Betrug*.

The trials clearly made their mark in the minds of alchemical authors. The language they used to denounce *Betrüger* sharpened, and gallows and punishments began to populate alchemical texts more frequently. Once false alchemists were merely sophists, impostors, or *Betrüger*; now they were "these knaves eminently worthy of the gallows."[35] In the foreword to his *Alchemia* of 1597 (the same year that Honauer was executed), Andreas Libavius explained that he hoped his readers would learn to "judge for yourselves, whether the essence, tincture, and so on, really is what it is acclaimed to be, or whether it is just a pure swindle."[36] While Libavius worried that publicizing alchemical knowledge could encourage further fraud, he explained that he had decided to publish his volume anyway because the state could easily defeat *Betrug:* "But where fraudulence is punished as it should be and the authorities are vigilant enough in their duties that scrupulous assayers are certified through examinations [*examina*] and a uniform standard for the examination can be deployed, then there is no reason to be apprehensive in this respect."[37] In Libavius's law-and-order approach to *Betrug,* a twin governmental strategy of testing and licensing true alchemists while punishing alchemical *Betrüger* would go a long way to sorting out the practice of alchemy. In this idea, at least, Libavius was far from original, as this is precisely what princely patrons of alchemy had been doing for several decades.

One of the most striking resonances of the alchemical fraud trials appeared among the original cluster of Rosicrucian texts, in the *Chymische Hochzeit Christiani Rosencreutz* (*The Chemical Wedding of Christian Rosenkreutz,* written ca. 1607–9, published 1616).[38] The likely author of the *Chymische Hochzeit,* the Lutheran utopian reformer Johann Valentin Andreae, inserted a striking scene in this alchemical parable that may well have been inspired directly by his close connections to the Württemberg court during Duke Friedrich's reign.[39] The *Chymische Hochzeit* begins when the humble pilgrim Christian Rosencreutz receives a mysterious invitation to a wedding, which serves as a metaphor for the union of opposites that results in the philosophers' stone. After a harrowing journey, he eventually arrives at a castle and is treated to a sumptuous banquet with the other guests, most of whom behave rather abominably. At the end of the evening, a Virgin appears with great fanfare to announce that before the wedding, the guests will all be weighed in order to determine whether they have received "the proper gifts from God" and therefore truly deserve to be among the guests who will witness the chemical wedding to come. "So that no rogue may do business here," she proclaims, "No knave slip in among the rest, / That all of you unhindered may / Enjoy the wedding undefiled, / Tomorrow every one of you / Upon the balance will be weighed."[40] With a warning that "he who dares beyond his powers / Would have done better not to come," the Virgin offers the guests a last chance to opt out of the weighing and leave the next morning. While those who are confident that they belong are then taken off to their rooms for the evening, the remaining few, whose consciences are pricking them and who have therefore decided not to risk the weighing, spend a rather unpleasant night tied up in the main hall.

The *Ceremoniis suspensionis ponderum* (ceremony of the suspension of the weights) follows the next day, ushering in a surprising turn of events. The Virgin praises the poor prisoners (Rosencreutz among them) who had opted to spend the night in the hall in lieu of the weighing, declaring, "That some of you are aware of your wretchedness pleases greatly my severe Lord, and he wishes you to be recompensed." As she ushers them off to the side to view the others' weighing, she tells them, "It may go better for you than for these daredevils who are still free." As the remaining guests are weighed one by one, emperor, nobleman, scholar, and "those pious gentlemen the arch swindlers [*Landbetriegern*]" alike find that they are too light for the scale; failing the test, they are immediately placed under arrest. "So few of the great crowd remained that I would be embarrassed to tell their number," Rosencreutz confesses. The Virgin decides that Rosencreutz and the other "poor coupled hounds" who had opted out of the weighing the night before should ascend the scale as well, but "without any danger to them, and only in good humor. There may be some worth among them."[41] While the new prisoners are taken away, those virtuous guests who passed the weighing then consider a punishment for the rest, with the Virgin now presiding as President.

After the Virgin and her "Senate" determine the proper punishment for the impostor guests' transgressions, the prisoners return to the hall. The Virgin's indictments and sentences serve as Andreae's vehicle for a commentary on the problem of *Betrug* at the beginning of the seventeenth century. Dividing into two groups those who had not passed the original weighing, the Virgin-President addresses one group:

> You should confess that you have lent credence too readily to false and spurious books; that you have thought too much of yourselves, and entered this castle to which no one invited you. If perhaps the majority of you simply wanted to slip in here in order to live more grandly and pleasurably, you have egged one another on to such outrage and effrontery that you have well deserved to suffer condign punishment.

After this group "humbly acknowledged it and raised their hands," the Virgin-President then turns to the other group:

> You know very well, and stand convicted by your consciences, that you have forged false and spurious books, fooled and swindled others, thereby lowering the royal dignity in everyone's eyes. . . . Now it has come to light what tricks you have played on the rightful guests, and

> how you have introduced unsuspecting ones. Everyone knows that you have been involved in open whoring, adultery, debauchery, and every kind of uncleanness, all of it in defiance of the well-known law of our kingdom. In short, you know that you have demeaned His Majesty before the common people; wherefore you should admit that you are publicly convicted arrant swindlers, toadies, and scoundrels [*Landbetrieger, Lecker und Buben*], who deserve to be sundered from decent folk and severely punished.[42]

This group then reluctantly confesses, in essence, to moral degeneracy and *Betrug*.

By separating out into two groups those who failed the test, the *Chymische Hochzeit* distinguished between the mere dupes and the active deceivers. The text clearly condemned them both for "demeaning" alchemy before the common people and using it as a selfish instrument for the generation of wealth rather than for what Andreae saw as its true purpose: the spiritual and medical benefit of humanity. (Certainly one can detect echoes of Brant's *Narrenschiff* here.) Even after confessing to its crimes, in fact, each group argues for a mitigated sentence by blaming the other: the first group places the burden on the second for leading them astray through deceit, while the second group faults the first for offering so much money in exchange for alchemical secrets that "each man had used his wits to get some. . . . Their books, moreover, had sold so well that anyone who could not live by other means was compelled to practice such a fraud."[43] And yet, as the Virgin-President's subsequent sentence makes clear, Andreae differentiated between those who actually committed acts of deceit and those who egged them on. The noblemen and rulers leave the castle unpunished, but with copies of a "*Catalogum Haereticorum oder Index expurgatorium* [Catalog of heretical works or list of works to be purged], whereby you may more wisely sift the good from the bad." The prisoners of less exalted rank are punished according to the severity of their crimes (as determined in the weighing earlier in the day). While some were stripped naked or branded and banished, the worst punishment was reserved for the *Landbetrieger*. "The arrant swindlers shall suffer capital or corporal punishment according to their deserts," the Virgin-President decrees, "with the sword, the rope, with water, or with rods. And this sentence shall be executed strictly as an example to all."[44] Watching the executions, Rosencreutz observes, "My eyes truly ran over at this execution, not so much because of the punishments, which they had well deserved for their crimes, but in reflection on human delusion."[45]

Andreae's judgment in the *Chymische Hochzeit* was clear. If princely patrons bore responsibility for the proliferation of *Betrüger*, princes themselves were redeemable and could be employed in eliminating *Betrug* altogether. Expert alchemists like Andreae simply needed to help them differentiate among the growing list of alchemical books, and, presumably, practitioners so that they supported only the worthy. The *Betrüger*, on the other hand, were clearly guilty in his eyes of moral degeneracy, which went hand in hand with alchemical deceit; they had to executed, both as a punishment for their egregious crimes and as a deterrent to others who might be considering taking a step down the path of alchemical *Betrug*. If Andreae did not witness Honauer's execution firsthand, one can only imagine he would have approved, given the striking judicial metaphor he chose to make his point about the damage false alchemists did to the "true" alchemy.[46]

The connections that Andreae made between weak morals and alchemical misdeeds highlight the difference between the way that princes approached *Betrug* (as breach of contract) and the much larger constellation of issues alchemical authors attached to the crime. For alchemical authors, the trials raised two important issues that went beyond whether alchemists had broken the law. One set of concerns was long-standing, namely alchemists' fear that an indiscriminating public would turn against all alchemy simply because of a few bad apples. When the trials questioned the reputation of both alchemy and alchemists anew, alchemists redoubled their efforts in print to distance themselves and their art intellectually and morally from the activities of the *Betrüger*. The trials also raised a second issue regarding the growing commercialization of alchemy. Sebastian Brant, of course, had already problematized the alchemical marketplace in the *Narrenschiff* of 1494, albeit at a time when alchemists were hardly hawking their wares openly alongside the vintners in the marketplace. A hundred years later, alchemical expertise was still for sale in an increasing variety of forms, but the largest, most public buyers were the princes; they were the ones who built laboratories, hired alchemists, and extended huge rewards to talented alchemists. It was the state's connection to entrepreneurial alchemy that drew public attention to the art, even as it raised truly fundamental questions about how alchemy operated in a commercial context.

Like Andreae, alchemical authors who took up these questions continued to narrate from the position of the alchemist-expert: they would guide would-be patrons away from the *Betrüger* and onto the path of proper alchemy. As Michael Maier insisted in his 1617 *Examen fucorum pseudo-chymicorum,* the trick was to separate out the "rotten drones" (or *Betrüger*), as he put it, from the "proper bees" (or true adepts). As he contended, "prudent is the person who distinguishes between good and bad, legitimate and illegitimate, true and false."[47] The difference was subtle, he confessed. "At first glance these drones have much in common with the proper bees or skilled chymists [*artificibus chymicis*]," Maier admitted, "such that one can hardly distinguish them from one another by their 'buzzing' or the din of their voices." Nevertheless, he felt that the discerning eye could see that the difference between the bees and drones was "even more than a thing from its opposite, or black from white." Maier's goal was to teach this acumen so that "the children of knowledge" could arm themselves against the *Betrüger* "with the acute understanding of an Odysseus before the harpies and sirens." This discernment could come only by studying the evidence of past fraudulent activity, drawn "in part from the books published by others, in part from the common gossip of the people and the complaints of victims." Drawing again on a judicial metaphor, Maier continued: "It will not be difficult to establish a verdict and a sentence for the many other means of deception that have yet to be recorded, but which will be brought out by them in days to come. Whoever you are, gentle reader, you will . . . distinguish the most divine of all arts, chymia, from the shams or drones and nonsense of impostors."[48] Maier hoped that his readers would eventually learn to be immune to the seductive "buzzing" of the deceitful *Betrüger* and stay on the proper alchemical path.

In case there was any doubt, the frontispiece of the *Examen fucorum pseudo-chymicorum* made it clear that Maier targeted both wealthy patrons and the criminal alchemists, the drones whose hostile attack on the true alchemy demanded an equally aggressive counterattack (fig. 13). On the left of the image, a humbly dressed true *chymicus* labors at a furnace, upon which stands an owl, a symbol of wisdom. Three elegantly clad men approach carrying alchemical vessels and, tellingly, a bag of coins. These figures, of course, represent the princely patrons who dangle money before alchemical practitioners and (as Andreae argued) thereby encourage alchemical fraud. To the right is no normal beehive with bees entering and exiting as they go about their work. Instead, this hive is under attack by a swarm of drones aimed straight at the entryway. Drawing on Aristotle's description of vicious conflicts that could arise between worker bees and drones, Maier's epigram accompanying the image praises the brave worker bees, who defend themselves against this attack of lazy but nevertheless aggressive drones and fight them to the death. Like these drones who seek

EXAMEN

FVCORVM

PSEVDO-CHYMICO-

RVMDETECTORVM

ET

IN GRATIAM VERITATIS AMAN-

tium ſuccincte refutatorum,

AVTHORE

MICHAELE MAIERO,

Com. Pal. Eq. Ex. Med. D.

FRANCOFVRTI

Typis Nicolai Hoffmanni, ſumptibus Theodori de Brij,

Anno M.CDXVII.

Figure 13. Frontispiece to Michael Maier, *Examen fucorum pseudo-chymicorum* (Frankfurt am Main: Theodor de Bry, 1617). (By courtesy of the Department of Special Collections, Memorial Library, University of Wisconsin–Madison.)

to steal the worker bees' hard-earned honey, Maier explained, the *Betrüger* live by attacking others and stealing the profits of their labors.[49]

Like princely patrons, Maier recognized a difference between the incompetent alchemists and those who were deliberately deceitful. The former were the *pseudo-chymici,* marked by their dangerous combination of ignorance and arrogance, and the latter were the drones, who cynically and deliberately used alchemy to con people out of their money. "I wish to call drones not those who simply err in their art or who are deceived by others," Maier explained, "but rather those who assault, attack, ensnare, and swindle others, simply with a mind to their own advantage and the others' detriment. Because these people are exceedingly harmful both to the *res publica* and to *chymia,* they must be immediately identified and guarded against."[50]

The point of the *Examen fucorum pseudo-chymicorum* was not only to educate, however, but also to condemn those who were derailing alchemy from its proper goals. Like the worker bees who fought back with a vengeance, Maier assailed the "pseudo-chemical drones" for their entrepreneurial orientation, ostentation, deceitful tricks, and lack of education. Some of Maier's critiques were hardly original. Echoing many earlier alchemical authors, for example, he argued that the drones had to rely on deceit to compensate for their lack of knowledge and training.[51] Maier's critique of entrepreneurial alchemy and its ostentation, however, was more innovative. Responding to the entrepreneurial practices that the fraud trials publicized, he argued for the absurdity of the trade in alchemical secrets. "It goes against all reason that someone who really had really mastered this great art, tested over and over again in experiments, would want to sell this knowledge to another for a piece of bread or a bit of gold." Maier's strategy focused on the difficulty of assigning commercial value to alchemical arcana, directly challenging princes who evaluated alchemy in these terms. "It is an unmistakable sign of the *pseudo-chymicus* that he wants to sell gold for gold something uncertain as fact and something priceless for very little," Maier reasoned: "If he doesn't really possess it [i.e., the alchemical secret], then it is as if he had sold wind and empty words for money. If the latter is the case, then the scoundrel receives too much money for the wind, and the buyer is cheated. If the former is the case, then the seller is cheated."[52] All of this added up to Maier's belief that true alchemy was not something that one could trade in the marketplace since it was, quite literally, priceless.

Not only did Maier inveigh against the very notion that alchemical knowledge could be commodified, but he also sought to damage the credibility of entrepreneurial alchemists by linking them inextricably with deception, impersonation, and moral depravity. Like princes, Maier construed this cluster of offenses as *Betrug*; he saw it, however, not as a narrow case of breach of contract, but as a much broader set of moral transgressions. Honauer's case, of course, seemed to exemplify the point with surprising clarity: impersonation and entrepreneurship went hand in hand. As a contemporary broadsheet commemorating his execution told the story,

> From Moravia he came this way
> Seeking to deceive
> He professed a grand art
> So everyone nearly believed.
> He pretended he wanted to make
> Out of iron nothing but pure gold:
> He appeared to be a great lord
> But he was nothing but a rogue.

Honauer's impersonation of a Moravian nobleman at the Stuttgart court was simply an extension of his masquerade as an alchemist (or vice versa), all of which was driven by the desire for profit. Maier elaborated on this notion with characteristic vitriol: "The '*Recipe*' of these frauds sounds like nothing but '*Decipe*,' and their knowledge is a procuress and their chemical process or art a whore. [The procuress] prostitutes the whore in order to gain some profit from her body, but what kind of body is it? Luscious, natural, and healthy? Absolutely not! To the contrary, not only is it deformed by ulcers, makeup, warts, and poisonous contagions, but it is also pestilential, which is to say that this type of learning teaches nothing but loss, this art nothing but anguish, this *alchymia* [or *Betrug*] nothing but laments and the gnashing of teeth."[53] Maier's description of false alchemy as a painted whore recalls Brant's alchemists and vintners, who used deception to disguise the shoddy quality of their wares.

Maier took Brant's notion a step further, however, by arguing that false alchemists were as likely to disguise themselves as they were to disguise their wares. Repeating a stock image of the alchemical dandy, Maier's *Examen fucorum pseudo-chymicorum* claimed that "these 'gold hunters' possess many coins,[54] ornate and expensive clothes, golden and silver instruments, rings, chains, numerous servants, and in their laboratory, which is outfitted with all the best types of equipment, they have artful ovens in great number."[55] Moreover, Maier claimed, *pseudo-chymici* amplified their pretension by freely boasting of their supposed talents to any willing ear.[56] Andreae used the metaphor of the mask in his 1616 *Chymische Hachzeit Christiani Rosenkroutz* to describe the same kind of impersonation. Observing the false guests who would later be exposed in the ceremony of the weights, Rosencreutz notes their foolish bravado and boasting at dinner the first night. Rosencreutz's companion, sharing his outrage, predicts these braggarts' downfall. "Just look at that toady," he remarks to Rosencreutz, "with what crazy talk and idiotic thoughts he seduces others to him. And that one there, who fools people with words of wonder and mystery. Believe me, though, the time will come when the masks in this masquerade will be ripped off, and the whole world will know what swindlers lurk beneath."[57]

Andrea's invocation of the mask goes to the heart of his and Maier's critique of the *Betrüger.* In essence, these critics took the performativity that was at the heart of *both* the alchemical persona *and* the marketplace in the early modern period and stood it on its head. As I have argued, the alchemical persona served a productive function in the early modern period by making the figure of the alchemist socially and culturally viable. Alchemists enacted this persona in all of their dealings with potential patrons by writing patronage letters, performing demonstrations, describing their recipes, and so on; all of these practices helped both parties establish the alchemist's legitimacy in an otherwise unstructured field and thus made it possible to put alchemy into practice. Maier, Andreae, and other critics, however, reconfigured these practices entirely by arguing that the enactment of the persona exposed alchemy to abuse. At best, these authors argued, such elements of alchemical practice missed the point by misguidedly highlighting the superficial performative aspects of alchemy—ostentatious apparatus, laborious processes, and flashy operations—rather than its fundamental elements. The *Betrüger* "favor ceremony more than the thing itself," Maier wrote; "I refer to as 'ceremony' those things which pertain to only the preparations, not to the essence of the thing."[58] At worst, these practices were nothing but the tricks of the confidence man: ruses, impersonations, deceits, and false performances designed to gain trust in order to perpetrate further frauds. (One is reminded here of Erasmus's *Betrüger* priest, who held Balbinus's interest with a series of processes and purchases without ever actually producing anything.) Maier, Andreae, and others, in other words, argued that all of the things that had become a necessary part of putting alchemy into practice as a public, collaborative activity were nothing but impersonation, lies, moral corruption, and *Betrug.*

In contrast, the true alchemist eschewed this kind of impersonation and promotion altogether, preferring alchemy as a private, solitary practice. The true *chymicus* was to be pious, modest, and secretive about his insights, like the humble Christian Rosencreutz.[59] Real alchemists, Maier wrote, "should be honest and upright men, both in public and in private, of good intentions and conscientious; that is, they should have not feigned, but true spirit of the art, acquiring it neither for the abuse of God, nor for the harm of one's fellow man."[60] Moreover, he continued, "a true goldmaker [*aurifex*] who knew how to complete his work in part or entirely would keep it all to himself. He must keep these things secret, so that others find out as little as possible about them and so that he can live in peace with his art, without the envy or suspicious looks of his neighbors."[61]

Maier viewed the bravado and bluster of the false alchemists not only as a giveaway, but also as morally objectionable. The impostors' boasting alone smacked of the deadliest of the seven deadly sins, pride. Moreover, because Maier assumed that would-be alchemists who decked themselves out as true practitioners did so deliberately in order to deceive, he accused them of a more general moral degeneracy. "Not the least of their moral failings is how they delight in lies, pride themselves on their own skill, and make fools out of others," he wrote. "This triple vice is like a hydra with many heads: if one cuts off one of them, another quickly grows in its place, and indeed more still in their place; in other words, it is the mother and propagator of many other, worse, evils."[62]

For alchemical authors like Maier, Andreae, and Libavius, therefore, the *Betrug* for which alchemists were convicted and even executed in the years around 1600 raised issues far beyond contracts and commitments. Whereas princes and legal scholars focused more narrowly on whether alchemists fulfilled their legal obligations or violated them, alchemical authors had larger issues in mind. Paradoxically, they worried about both alchemy's successes and its failures. One the one hand, authors like Maier worried about alchemy's declining reputation, particularly the way in which the very public failure of alchemy, symbolized by the executions, could turn people against what they believed to be a noble art. On the other hand, the trials publicized the extent to which a productive, entrepreneurial alchemy had taken root in the Holy Roman Empire and was increasingly convincing to princely patrons. For alchemists like Maier, this kind of alchemy was a problem not only because it ignored or minimized alchemy's philosophical and spiritual benefits; it also created serious competition in the patronage marketplace. Maier, after all, was competing with alchemists like Georg Honauer for princely patronage, and he feared that he might lose out as princes doled out ever-higher sums to rivals who promised exaggerated financial winnings.[63] From this perspective, alchemy's success was its weakness, as it was diluted and distorted beyond recognition.

Subsequent authors would play up the trope of defending alchemy by condemning false alchemists who peddled their deceits in books, if not in person. In 1635, for example, a short anonymous treatise appeared, entitled *Via veritatis, das ist: Ein warhafftiger philosophischer Bericht, aus dem rechten unnd wahren Fundament der Natur genommen, und den irrenden Alchimisten, so jhre gute Pfeil verschossen, und durch die falsche Process-bücher in grosse Armuth gerathen, zum besten (damit sie einmal uff den rechten Weg gebracht, und sich jhres Schadens erholen können)* (True path; or, A true philosophical report, taken from the proper and true fundament of nature, for the benefit of the erring alchemists, who shoot their good arrows crookedly, and who are guided into great poverty with the false books of processes, so that they may be brought onto the proper path at once and can recuperate their losses). The *Via veritatis* was typical of the many texts to turn to print to defend alchemy against those whose observations, and perhaps even personal experience, had led them to a negative conclusion. "Dear gentle reader," the preface began, "for ages now and even more so in these wretched and distressing times, many distinguished people, both of high and low station, have amused themselves with chymical studies, with the enduring hope that they will thereby produce a livelihood for themselves and theirs; most of them, however, have threshed nothing more than empty straw, and others have achieved nothing but great misery, woe, and, finally, extreme poverty and reduction to beggary, as can be shown with many examples. What is more, since they believe they have been deceived, they are not afraid to speak abusively of the art, and to topple the true philosophers together with all of their writings from their heights."[64] The *Via veritatis* went on to defend alchemy's virtue, predictably using the false alchemists and their books as a foil for articulating the true art. What is in a sense most interesting about works that condemned false alchemists in the service of the true alchemy is that their authors directed them to the reading public. Despite critiques of entrepreneurial alchemists for focusing too much on superficial matters like reputation, these authors evidently cared about such issues as well, or else they would not have gone to such lengths to put their rebuttals in print. They did so in part to attract the support of patrons, of course, but also because they, too, recognized that authority in alchemical matters was determined, in part, through performances of expertise in the marketplace.

Deliberate Trickery?

So were alchemical drones, in fact, storming the hive of proper alchemy in seventeenth-century central Europe? Obviously Maier thought so, and, lest there be any doubt, he concluded the *Examen fucorum pseudochymicorum* with a list of the various tricks the swindlers allegedly used. Maier did not claim to have firsthand experience with such sleights of hand, but rather gleaned most of them from Khunrath's 1597 "Treuhertzige Warnungs-Vermahnung" (Heartfelt warning and admonition).[65] This *Betrüger* literature from people like Maier and Khunrath, in essence, functioned as a parallel genre to the stories of successful transmutations, known as transmutation histories, that increasingly appeared in seventeenth-century Europe. Just as transmutation histories often served to bolster early modern alchemists' belief in the powers and reality of the philosophers' stone, so too did stories of alchemical fraud serve to authenticate the existence of *Betrüger.* Stories of true and false alchemy, in other words, emerged in tandem.[66] Maier assumed that Khunrath's report was largely accurate and that "these frauds and deceptions [either] have been used at some point by pseudo-chymists or could be used by impostors."[67] Citing Khunrath's text, point by point, Maier described several alchemical processes that are obviously deliberate deceptions, by both sixteenth-century and twenty-first century standards. For example, Maier, via Khunrath, cites the use of double-bottomed crucibles, in which a thin false bottom conceals hidden gold. Using a small iron hook, the false alchemist could pierce the false bottom during laboratory operations, thus releasing the gold into the crucible.[68] In other cases, however, the techniques Maier discusses are less clearly deceptive. The *vitriolum album* he describes, for example, supposedly crystallizes into gems that look like rubies.[69] For Maier this represents a misleading process, resulting in a counterfeit gemstone rather than a real one. Other alchemists, however, did not see this as misleading or counterfeit, but rather as a valuable recipe for the human manufacture of gems.[70] Sömmering's colleague Anna Zieglerin, for example, discussed precisely this kind of "white vitriol" in the alchemical recipe book she dedicated to Duke Julius in 1573, clearly representing the product as a manufactured gemstone nonetheless, rather than one simply found in nature.[71] What some practitioners defined as legitimate practical alchemy, in other words, others defined as a trick.

We will never be entirely sure whether or not alchemists made use of such tricks, and we will always be even more uncertain about what they were thinking when they did so. Most alchemical activity, of course, is lost to us today because it was never documented, but we can evaluate the evidence that has survived. On their own, the *Examen fucorum pseudo-chymicorum* and the "Warnungs-Vermahnung" are deeply problematic as evidence that such manipulations actually occurred. The authors of these texts had a stake in propagating stories of fraud, whether or not such reports were true; stories of *Betrug* allowed them to increase their own authority as alchemist-experts. By damaging the credibility of entrepreneurial alchemists, alchemists like Maier could increase their own authority in a competitive patronage marketplace.[72] Furthermore, neither Maier nor Khunrath claimed to have firsthand knowledge of such tricks; both simply reported what they gathered from other texts and from rumors. These alchemical authors' own motivations should give us pause before we accept uncritically the evidence of fraud they offer. At a minimum, we ought to approach these stories of fraud with the same critical eye that we so easily turn on stories of successful transmutations; why, after all, should stories of alchemical tricks be any more believable to us today than stories of transmutation?

If evidence of the actual use of deliberately deceptive alchemical tricks is what we are after, we can learn much more by getting away from polemical texts like Khunrath's and Maier's and looking at alchemy in practice. If in fact such alchemical legerdemain was common, we might expect it to have surfaced in the context of legal trials against alchemists. Authorities explicitly sought out evidence of *Betrug* during these trials, after all, and interrogators had recourse to torture to elicit confessions where necessary. The documentation generated by these trials, of course, is problematic in other ways, and we must approach it carefully. But let us begin with the most basic question: did alchemical fraud trials generate confessions to alchemical tricks? Interestingly, with the notable execution of Nüschler's admission of slipping a bit of gold in during his transmutations, the alchemical fraud trials do not reveal evidence of the kind of fraudulent manipulation that Maier and others report; the issue simply did not come up in trials because this was not the kind of *Betrug* that princes were focused on. Alchemists skipped town and failed to fulfill their contracts, to be sure, but there is little evidence that they relied on sleights of hand to swindle their princely patrons. Princes did not accuse alchemists of such crimes, nor did alchemists confess to them.

What princes and alchemists alike *were* interested in, however, was the issue of intent: did alchemists *intend* to deceive their patrons when they arrived at court, if not through deceptive transmutations, then through schemes to take the money and run? Because they involve individuals' mind-sets and inner thoughts, intentions are extremely difficult for the historian to recover, even with the best and most direct of sources. In this case, because the alleged *Betrüger* themselves did not publish their own views, the best sources the archives offer are their trial dossiers: records of interrogations (often conducted under torture) and pardon pleas. These are problematic sources indeed, but in many cases they are all that we have. What, then, do the sources tell us on this point? Unlike alchemical tricks, the issue of intent did come up explicitly in interrogations or pardon pleas. Philipp Sömmering, for example, insisted on his good will, claiming that "he had definitely heard of the strict punishment by [his patron Duke Julius]. . . . However he had nothing untrue in his heart against [the duke]."[73] In other cases, the evidence is less direct. The confession of the Cypriot alchemist Marco Bragadino lists the nine times he accepted money to perform alchemical processes but fled before carrying them out. Interestingly, Bragadino was silent on the issue of intent, although his interrogators clearly asked. After each count of debt, the confession reads, "He did not say whether he intended to steal it or repay it."[74] Bragadino did not deny that he had broken his agreements, in other words, but refused to admit or deny that he had intentionally tricked his benefactors. In these cases legal officials were clearly looking for evidence of deliberate swindles, often employing torture to extract a confession of guilt, and yet practitioners *still* refused to admit deceitful intentions. On the issue of intent, the trials offer denials, but no confessions to premeditated breach of contract.

Not surprisingly, alchemists on trial for failing to fulfill their contracts often cited fear as a strong motivation for their actions. They claimed to have signed contracts with the best of intentions, fully believing that they would be able to deliver what they promised, only to find later that their process would not work or that they needed more time. In 1597, for example, Georg Honauer appealed all the way to the Holy Roman Emperor Rudolf II in a plea for his life. The alchemist explained that he had been delayed in his alchemical work for Duke Friedrich because of his less-than-capable assistants. Friedrich was understandably upset by this delay and, according to Honauer, threatened to "lay a hand on me." At the same time, Honauer continued, "trustworthy people" warned him to be careful, which caused him "great injury and great terror." It was this fear, Honauer claimed, that led him to flee suddenly—not a preplanned scheme to take money from Friedrich.[75] In his plea for mercy, Hans Heinrich Nüschler made a similar claim. Nüschler stated that he slipped a bit of gold into his transmutation out of fear only upon discovering that his transmutory powder was useless; he denied that he set out for Stuttgart with the intent to involve the duke in an elaborate con game.

So what, then, can we conclude from the alchemical fraud trials about the practice of deceit? First, the kinds of tricks that Khunrath and Maier described largely did not come up; Nüschler's case was exceptional in this regard. What princely prosecutors were interested in was something else: whether alchemists who broke their contracts did so as part of a premeditated plan or only as an act of desperation. On this point, the alchemists either were silent or insisted that they acted out of desperation or fear alone. We could interpret this evidence in a variety of ways. We might choose to believe Nüschler, for instance. After all, he surely knew of Duke Friedrich's execution of Honauer in 1597 and Petrus Montanus in 1600; the gallows was still standing in the center of Stuttgart when he arrived, making the consequences of alchemical fraud all too clear. We could argue that, knowing this risk, it is unlikely that he would have proposed his process to the duke without actually believing in his own skill. On the other hand, since Nüschler did confess to deceit, it is equally plausible that it was at the heart of his strategy from the moment he arrived in Stuttgart; perhaps he heard of Friedrich's interest in alchemy, and decided to gamble on winning Friedrich's favor and largess by performing a false transmutation. The fact is, given the extant evidence, either interpretation is plausible. Ultimately, the evidence does not reveal what Nüschler was thinking, and thus we must admit its limits and leave the question of intent unresolved rather than making arguments based on assumptions. We cannot condemn the so-called frauds for more than their patrons did: for breaking contracts that promised far more than the alchemist could produce.

In one sense, what the *Betrug* trials in the years around 1600 prove is actually surprisingly limited: the trials were about contracts, not about cons (let alone demons, magic, or whatever else we might assume). Quite simply, some alchemists could not fulfill the contracts they had signed, and when they tried to escape their responsibilities rather than confess failure, they paid the highest price. This rather modest conclusion, however, belies the significance of the trials for the history of alchemy. They demonstrate the logical consequences of the kind of entrepreneurial alchemy that had developed among the princely courts of the Holy Roman Empire; when patrons and practitioners held alchemy to a precise, quantifiable standard and expected an almost industrial level of productivity, it ultimately failed. More specifically, contracts structured risk so that individual alchemists bore the responsibility for alchemy's failure, sometimes in the most dramatic fashion. This transference of risk and failure to individual alchemists, strangely enough, proved to be advantageous to alchemical authors as well as princes. By blaming alchemy's most public failures on individual practitioners, alchemical authors and princes alike found a convenient explanation for what had gone wrong. Princes had simply placed their trust in the wrong kinds of practitioners, who managed to conceal their corrupt morality or incompetence with their formidable skills of impersonation.

Paradoxically, then, despite the differences in the ways that princes and alchemical authors responded to alchemical fraud trials, the very notion of a "false" alchemist played a central role in the continuing belief in alchemy's promise as a productive art in early modern Europe. For the definition of some practitioners as frauds or impostors implied that somewhere true alchemists existed; the hopeful patron simply had not yet found one. As Duke Friedrich explained, "we have found from particular experience that in no other art on earth is there more fraud and falsehood [*betrugs und falschs*] than in the supposed art of the sophistic goldmaker who goes around with a tincture that he himself did not even make. . . . [It is] nevertheless undeniable that the ancient sages had a genuine natural medicine, with which they not only defended the human body against all illnesses until the predetermined hour of death, but [also] transmuted and changed imperfect metals into gold of the highest quality, as we have not only found in our good philosophical books [but also] have seen with our own eyes."[76] In the end, it was probably much easier for all to believe that failed alchemists were deliberate frauds or incompetent than it was to accept that they had honestly and knowledgeably tried to carry out alchemy's promise and still had failed. After all, patrons were as heavily invested in the alchemical dream as a panacea for a troubled economy as alchemists themselves were invested in its social, intellectual, and material potential. As long as alchemy's failures could be attributed to an individual practitioner's duplicity or incompetence, then the dream of alchemical productivity would persist and early modern princes and alchemists alike would continue their efforts to realize it.

CONCLUSION

The Problem of Authority

On one level, this has been a book about a new kind of entrepreneurial alchemical practice that emerged in the sixteenth century around the princely courts of central Europe. In this context, alchemy was integrated into territorial rulers' long-standing interest in the economic vitality of their lands. In the late fifteenth and early sixteenth centuries, these princes directly invested in mining enterprises, financing technological innovation and centralizing control over these crucial industries as part of a broader process of state building. In the sixteenth century, they incorporated alchemy into these projects, employing alchemists to pursue new processes for refining and producing precious metals. Patrons and practitioners drew on the structures not only of patronage, but also of the mining trade in order to put alchemy into practice in this context. Thus princes signed contracts with alchemists, just as they had previously signed investment contracts with mining companies, outlining specific processes, expected profits, and financial responsibilities of both parties; they also built large laboratories and created elaborate organizational structures to regulate them. This kind of alchemical practice was "entrepreneurial" both for the princes, who hoped it could revive or extend the failing mines that were the lifeblood of central European territories, and for the alchemists, who hoped to attract princely investment in expensive alchemical work and, perhaps, earn substantial rewards for successful alchemical projects. This entrepreneurial turn in the practice of alchemy had important consequences, for the structures that patrons and alchemists created held alchemists to very specific standards of profit and productivity that they could not always meet. The fraud trials around 1600 were the direct product of these entrepreneurial arrangements, which set up structures for alchemy that encouraged some alchemists to take on too much risk and to promise too much for their art and their abilities.

These entrepreneurial alchemical practices, however, were embedded in a much more complex understanding of the figure of the alchemist that emerged not only from princely courts, but also from the pens of cultural commentators and alchemical authors. Debates about alchemy in this period increasingly focused on the alchemist as a social type: what was distinctive about a real alchemist (or, for that matter, a false one), and what sort of social and cultural meanings could this figure bear? In the sixteenth century, three interrelated contexts simultaneously shaped this debate. Traditional alchemical texts presented the image of the alchemist as scholar, prophet, and artisan, a figure who combined divine revelation, study, and laboratory practice to produce both philosophical knowledge and useful products. At the same time, humanist observers increasingly deployed the alchemist as a potent symbol of deceit and fraud, a vehicle for critiquing early modern commerce and social mobility. Finally, entrepreneurial alchemical practices situated the alchemist in the context of early modern state building and political economy. These three contexts for alchemy—as ancient art, cultural symbol of deceit, and entrepreneurial practice—all vied to define the figure of the early modern alchemist. The focus of this book has oscillated among these perspectives, exploring how they interacted with and responded to one another in constructing a particularly early modern version of alchemical practice.

The concept of alchemical fraud, or *Betrug,* ran through every context for alchemy, shaping its practice and practitioners in crucial and complex ways. As I have endeavored to show, alchemical fraud was not a natural category; it is not somehow intrinsic to alchemy, any more than it is to physics, medicine, theology, or baking. Rather, *Betrug* was a specific product of alchemy's vitality and expansion at the end of the Middle Ages and into the seventeenth century, as well as a sign of the controversy this vitality engendered. As alchemy attracted new practitioners through print and vernacularization, as well as new support from princely patrons who were willing to invest money and laboratory space, it also attracted new audiences and commentaries. The figure of the alchemist leapt off the pages of Latin philosophical treatises and into the ships of fools, becoming a new social type whom central Europeans might actually imagine encountering in the cities and courts of the Holy Roman Empire. This new visibility of alchemy and its practitioners spurred a debate about fraud that always carried a double meaning. Accusations of *Betrug* could focus on *alchemical practices* and whether they were in fact philosophically grounded, effective, commodifiable, or deceptive; but they could also focus on *alchemical practitioners* and whether they were honestly and capably embodying the alchemist's persona or merely impersonating it to other nefarious ends. Debates about fraud in this double sense, then, were never just an inevitable condemnation of alchemical hangers-on, but a way to grapple with alchemy itself.

What is most striking about this debate, in the end, is the range of people it engaged, from humanist satirists and princes to alchemical authors and legal scholars. Alchemy, in other words, not only involved natural philosophers and others who pursued natural knowledge in early modern Europe; it resonated far beyond the realm of scholarship, touching on social mobility, state building, commerce, humanism, and even crime and punishment. Alchemy, therefore, is not merely a subject for historians of science or intellectual history, but a venue for exploring many of the central issues in early modern European history.

Alchemical fraud resonated widely because it was always about asserting authority. From Philipp Sömmering to Michael Maier, alchemists claimed authority as interpreters of alchemical tradition in part by condemning others' failure to do so properly. They made a bid for intellectual authority, hoping to establish their own credentials as upstanding alchemist-experts with special access to alchemy's secrets as well as the social status that attended their knowledge and expertise. In the process, they sought to draw the boundaries of their knowledge community where few other markers of those boundaries existed. For humanist critics and artists, meanwhile, less was at stake; alchemists were only one type of fool among the endless supply available to inspire their pens and brushes. Nevertheless, these social critics exploited alchemists to claim their own authority as cultural commentators and interpreters, challenging alchemists themselves for the right to decode the social and cultural meaning of alchemy for broad audiences. Princes, finally, asserted an entrepreneurial alchemical practice in the late sixteenth century that not only increased their authority as sovereigns in the political sphere, but also affirmed their determination to mold early modern natural knowledge to their own ends. The princes of the Holy Roman Empire were not alone in this regard, of course, but must be seen alongside other early modern sovereigns who employed natural knowledge in early modern state and empire building. In providing an opportunity for all of these various groups to assert their own authority, debates about fraud revealed how much was at stake with alchemy, whether in the intellectual, social, political, or economic realm.

The battle for alchemical authority at the heart of this book, therefore, was waged on many fronts and with many agendas and results. At the center of the battle, however, were always the *"Betrüger"* themselves, alchemists like Philipp Sömmering, Georg Honauer, Marco Bragadino, and the others who populate this book. Although these three met their deaths on the gallows in Wolfenbüttel, Stuttgart, and Munich, nevertheless, they left their mark by asserting their own claim to natural knowledge in the Holy Roman Empire. They did so—through their actions if not their words—in ways that may have been dismissed by or escaped the attention of historians, but these alchemists' boldness and confidence is still striking today. The European archives, in fact, are overflowing with sources that document the world of natural knowledge claimed and created by artisans, sailors, empirics, artists, priests, and noblewomen—a whole range of people beyond the well-known scholars who chose print as the venue for publicly asserting their claims to knowledge. We simply will not fully appreciate the scope of natural knowledge and its meanings in early modern Europe until we can incorporate the new sources and new subjects that the archives reveal. In the end, then, any study of natural knowledge in early modern Europe must reckon not only with people like Isaac Newton and Robert Boyle, but also with people like Philipp Sömmering.

NOTES

INTRODUCTION

1. The voluminous documentation of Sömmering and his accomplices' career at the court of Duke Julius of Braunschweig-Wolfenbüttel (1528–89) comprises thousands of pages and may be found in NStA Wolfenbüttel, 1 Alt 9, Nrs. 306–36. The principal secondary source for this case is Albert Rhamm, *Die betrüglichen Goldmacher am Hofe des Herzogs Julius von Braunschweig: Nach den Processakten* (Wolfenbüttel: Julius Zwißler, 1883). For a briefer summary of the case (based heavily on Rhamm's interpretation), see Prof. H. Wr., "Die Goldmacherbande am Hofe des Herzogs Julius von Braunschweig in Wolfenbüttel," *Niedersachsen* 14 (1908/1909): 346–51. I have taken these details from Sömmering's first statement in his trial, "Was Philip Sömmering anfangs und zum Eingang in den gute berichtet und aufgesagt," n.d. [June 1574], NStA Wolfenbüttel, 1 Alt 9, Nr. 311, fols. 14r–17v. On Sömmering's expertise in mining matters, see "Klagesartikeln gegen Philippen," n.d. [1574], NStA Wolfenbüttel, 1 Alt 9, Nr. 310. On the gun barrel, see Rhamm, *Die betrüglichen Goldmacher,* 15. On Sömmering's career as a pastor, see Rhamm, *Die betrüglichen Goldmacher,* 7; also Thüringisches Hauptstadtsarchiv Weimar, Ernestinisches Gesamtarchiv, Reg. P, fols. 167–68 F3, Nr. 4. Unless otherwise indicated, all translations are mine.

2. The contract itself is no longer extant, but Sömmering described it during his interrogation as follows: He explained "how he [Philipp] came to Wölfenbüttel again, [and] showed Illustrissimus [i.e., Julius] his letter from Duke Johan Friederich [of Sachsen-Gotha], because he wanted to enter the service of His Grace. He requested two thousand taler for the art and said that he wanted to do whatever he could for him [Julius], and promised His Grace [that he would make] one loth of tincture in a certain period of time or return the money to him again. Illustrissimus contracted with him for this, and had the money (as well as more later) given to him." "Was Philipp Sömmering anfangs." On Sömmering's arrangement with Julius, see also "Gütliche Aussage des Philipp Sömmering," 9 July 1574,

NStA Wolfenbüttel, 1 Alt 9, Nr. 308, fols. 49v–56; and "Verhör und Aussagen Ph. Sömmerings," 10 July 1574, NStA Wolfenbüttel, 1 Alt 9, Nr. 309, fol. 13v.

3. For the details of this conflict, known as the "Grumbachschen Händel," see Peter Elsel Starenko, "In Luther's Wake: Duke John Frederick II of Saxony, Angelic Prophecy, and the Gotha Rebellion of 1567" (PhD thesis, University of California, Berkeley, 2002); Jost Weyer, *Graf Wolfgang II. von Hohenlohe und die Alchemie: Alchemistische Studien in Schloß Weikersheim, 1587–1610,* ed. Historischen Verein für Württembergisch Franken, Stadtarchiv Schwäbisch Hall and Hohenlohe-Zentralarchiv Neuenstein, Forschungen aus Württembergisch Franken, vol. 39 (Sigmaringen: Jan Thorbecke Verlag, 1992), 310; and Rhamm, *Die betrüglichen Goldmacher,* 1–3.

4. Anna Maria Zieglerin to Herzog Julius, 12 December 1571, NStA Wolfenbüttel, 1 Alt 9, Nr. 306, fol. 4. On Zieglerin, see Tara E. Nummedal, "Alchemical Reproduction and the Career of Anna Maria Zieglerin." *Ambix* 48 (2001): 56–68. Zieglerin is also the subject of my next book, *Anna Zieglerin and the Lion's Blood: An Alchemist's Career in Reformation Europe.*

5. NStA Wolfenbüttel, 1 Alt 9, Nr. 306, fols. 99–93. Hoping to put these rumors to rest, Julius wrote to Philipp's former patron, Duke Johann Friedrich of Sachsen-Gotha, and asked him to vouch for Philipp's character, which the Saxon duke gladly did. See Rhamm, *Die betrüglichen Goldmacher,* 21; and NStA Wolfenbüttel, 1 Alt 9, Nr. 307, fols. 169–70.

6. Katharina, Markgravess zu Brandenburg, to Hedwig, Duchess von Braunschweig-Wolfenbüttel, 18 June 1573, NStA Wolfenbüttel, 1 Alt 9, Nr. 307, fols. 55–56.

7. On Kettwig (or Kettwich), see Hans-Joachim Kraschewski, *Wirtschaftspolitik im deutschen Territorialstaat des 16. Jahrhunderts: Herzog Julius von Braunschweig-Wolfenbüttel* (Cologne: Böhlau Verlag, 1978), 98; and Rhamm, *Die betrüglichen Goldmacher,* 28–29.

8. The Danish crown issued the warrant on behalf of Kettwig's former employer Heinrich von Ratznau (1526–99). Coincidentally, Heinrich Rantzau was also patron to Peter Maier, the father of the well-known alchemist, physician, and Rosicrucian Michael Maier. See Claus Priesner and Karin Figala, eds., *Alchemie: Lexikon einer hermetischen Wissenschaft* (Munich: Beck, 1998), 124. On Rantzau see Dieter Lohmeier, "Heinrich Rantzau und die Adelskultur der frühen Neuzeit," in *Arte et Marte: Studien zur Adelskultur des Barockzeitalters in Schweden, Dänemark und Schleswig-Holstein,* ed. D. Lohmeier (Neumünster: K. Wachholtz, 1978), 67–84.

9. Kettwig's colorful denunciation of "the godforsaken . . . and dishonorable, quartered, hanged, and vinegary rogues, whores, and knaves [*die Gott vergessene . . . und erlose, Viertheill henckige und brennessige schelmen huren und buben*] Philip Schreiber, born Sommerinck von Dambach, Anna Maria, and her husband, known as one Heinrich Schombach" can be found in NStA Wolfenbüttel, 1 Alt 9, Nr. 307, fols. 182–87. See also Rhamm, *Die betrüglichen Goldmacher,* 35.

10. For all of the documentation regarding Kettwig, see NStA Wolfenbüttel, 1 Alt 9, Nr. 318, fols. 1–190.

11. Johann Georg, Markgraf zu Brandenburg, to Julius, 4 February 1574, NStA Wolfenbüttel, 1 Alt 9, Nr. 307, fols. 90–91.

12. See the "Klagesartikel and Fragesartikel" in the trial dossier, n.d. [1574], NStA Wolfenbüttel, 1 Alt 9, Nr. 310, fols. 68–93v; and "Auszug der brandenburgischen Schöppensprüche gegen Ph. Sömmering, Heinrich Schombach, Anna Maria Ziegler, Jobst Kettwig, Sylvester Schulvermann, Dr. Georg Kommer, Bernd Hueffner, Hans Hoyer," n.d. [1574], NStA Wolfenbüttel, 1 Alt 9, Nr. 311, fols. 23–33. The original list of charges is no longer extant.

13. The Mühlentor was roughly where the Herzog-August-Bibliothek stands today.

14. Most prominently, Carlo Ginzburg, *The Cheese and the Worms: The Cosmos of a Sixteenth-Century Miller* (Baltimore: Johns Hopkins University Press, 1980); and Natalie Zemon Davis, *The Return of Martin Guerre* (Cambridge, Mass.: Harvard University Press, 1983). See also Edward Muir and Guido Ruggiero, *History from Crime* (Baltimore: Johns Hopkins University Press, 1994); Guido Ruggiero, *Binding Passions: Tales of Magic, Marriage, and Power at the End of the Renaissance* (Oxford: Oxford University Press, 1993), and "The Strange Death of Margarita Marcellini: *Male,* Signs, and the Everyday World of Pre-modern Medicine," *American Historical Review* 106, no. 4 (2001): 1141–58; Thomas V. Cohen and Elizabeth S. Cohen, *Words and Deeds in Renaissance Rome: Trials Before the Papal Magistrates* (Toronto: University of Toronto Press, 1993); Richard L. Kagan, *Lucrecia's Dreams: Politics and Prophecy in Sixteenth-Century Spain* (Berkeley: University of California Press, 1990); and Judith Brown, *Immodest Acts: The Life of a Lesbian Nun in Renaissance Italy* (New York: Oxford University Press, 1986).

15. Inquisition records, on the other hand, have been important sources for a number of historians of science and medicine. See, for example, David Gentilcore, *Healers and Healing in Early Modern Europe* (Manchester: Manchester University Press, 1998); Hilary Gatti, *Giordano Bruno and Renaissance Science* (Ithaca, N.Y.: Cornell University Press, 1999); Pietro Redondi, *Galileo Heretic* (Princeton, N.J.: Princeton University Press, 1987); and William Eamon, "Cannibalism and Contagion: Framing Syphilis in Counter-Reformation Italy," *Early Science and Medicine,* no. 1 (February 1998): 1–31, and "'With the Rules of Life and and Enema': Leonardo Fioravanti's Medical Primitivism," in *Renaissance and Revolution: Humanists, Scholars, Craftsmen, and Natural Philosophers in Early Modern Europe,* ed. J. V. Field and Frank A. J. L. James (Cambridge: Cambridge University Press, 1993), 29–44.

16. Natalie Zemon Davis, *Fiction in the Archives* (Stanford, Calif.: Stanford University Press, 1987), 25.

17. For two examples of the older view, see Henry Carrington Bolton, *The Follies of Science at the Court of Rudolph II, 1576–1612* (Milwaukee: Pharmaceutical Review Publishing Co., 1904); and Theodor Wagner, "Wissenschaftlicher Schwindel aus dem südlichen Böhmen," *Mittheilungen des Vereins für Geschichte der Deutschen in Böhmen* 16 (1878): 112–23. Classic works on alchemy's role in the

Scientific Revolution include Frances A. Yates, *Giordano Bruno and the Hermetic Tradition* (Chicago: University of Chicago Press, 1964); and Betty Jo Teeter Dobbs, *The Foundations of Newton's Alchemy or, "The Hunting of the Green Lyon"* (Cambridge: Cambridge University Press, 1975), and *The Janus Faces of Genius: The Role of Alchemy in Newton's Thought* (Cambridge: Cambridge University Press, 1991). More recent studies include William R. Newman, *Gehennical Fire: The Lives of George Starkey, an American Alchemist in the Scientific Revolution* (Cambridge, Mass.: Harvard University Press, 1994), and *Promethean Ambitions: Alchemy and the Quest to Perfect Nature* (Chicago: University of Chicago Press, 2004); Lawrence M. Principe, *The Aspiring Adept: Robert Boyle and His Alchemical Quest* (Princeton, N.J.: Princeton University Press, 1998); William R. Newman and Lawrence M. Principe, *Alchemy Tried in the Fire: Starkey, Boyle, and the Fate of Helmontian Chymistry* (Chicago: University of Chicago Press, 2002); Pamela H. Smith, *The Business of Alchemy: Science and Culture in the Holy Roman Empire* (Princeton, N.J.: Princeton University Press, 1994), and *The Body of the Artisan: Art and Experience in the Scientific Revolution* (Chicago: University of Chicago Press, 2004); Bruce T. Moran, *Distilling Knowledge: Alchemy, Chemistry, and the Scientific Revolution* (Cambridge, Mass.: Harvard University Press, 2005); and William R. Newman, *Atoms and Alchemy: Chymistry and the Experimental Origins of the Scientific Revolution* (Chicago: University of Chicago Press, 2006).

18. Bruce T. Moran, *The Alchemical World of the German Court: Occult Philosophy and Chemical Medicine in the Circle of Moritz of Hessen (1572–1632)* (Stuttgart: Franz Steiner Verlag, 1991), "German Prince-Practitioners: Aspects in the Development of Courtly Science," *Technology and Culture* 22, no. 2 (1981): 253–74, and Moran, ed., *Patronage and Institutions: Science, Technology, and Medicine at the European Court, 1500–1750* (Rochester, N.Y.: Boydell Press, 1991); Paula Findlen, *Possessing Nature: Museums, Collecting, and Scientific Culture in Early Modern Italy* (Berkeley: University of California Press, 1994), and "Courting Nature," in *Cultures of Natural History*, ed. Nicholas Jardine, James A. Secord, and E. C. Spary (Cambridge: Cambridge University Press, 1996), 57–74; Mario Biagioli, *Galileo Courtier: The Practice of Science in the Culture of Absolutism* (Chicago: University of Chicago Press, 1994); John Robert Christianson, *On Tycho's Island: Tycho Brahe and His Assistants (1570–1601)* (Cambridge: Cambridge University Press, 2000); and Helen Watanabe-O'Kelly, *Court Culture in Dresden: From Renaissance to Baroque* (Houndmills: Palgrave, 2002). For two recent exhibition catalogs on important courts for early modern natural knowledge, see Dirk Syndram and Antje Scherner, eds., *Princely Splendor: The Dresden Court, 1580–1620* (Dresden: Staatliche Kunstsammlungen Dresden; Milan: Mondadori Electa, 2004); and Eliška Fučíková, *Rudolf II and Prague: The Court and the City* (Prague: Prague Castle Administration; London: Thames and Hudson, 1997).

19. R. J. W. Evans, *Rudolf II and His World: A Study in Intellectual History* (Oxford: Oxford University Press, 1973); Pamela H. Smith, "Alchemy as a Language of Mediation at the Habsburg Court," *Isis* 85 (March 1994): 1–25, and *The*

Business of Alchemy; Bruce Moran, *The Alchemical World of the German Court*; Jost Weyer, *Graf Wolfgang II. von Hohenlohe und die Alchemie: Alchemistische Studien in Schloß Weikersheim, 1587–1610*, ed. Historischen Verein für Württembergisch Franken, Stadtarchiv Schwäbisch Hall and Hohenlohe-Zentralarchiv Neuenstein, Forschungen aus Württembergisch Franken, vol. 39 (Sigmaringen: Jan Thorbecke Verlag, 1992); Hans-Georg Hofacker, *". . . sonderlich hohe Künste und vortreffliche Geheimnis": Alchemie am Hof Herzog Friedrichs I. von Württemberg—1593 bis 1608.* (Stuttgart: Verein der Freunde des Chemischen Instituts Dr. Flad e. V., 1993).

20. Hereward Tilton, *The Quest for the Phoenix: Spiritual Alchemy and Rosicrucianism in the Work of Count Michael Maier (1569–1622)*, ed. Cristoph Markschies and Gerhard Mueller, Arbeiten zur Kirchengeschichte 88 (Berlin: Walter de Gruyter, 2003); Dobbs, *The Foundations of Newton's Alchemy*; Newman, *Gehennical Fire*; Principe, *The Aspiring Adept*; Smith, *The Business of Alchemy*.

21. Several recent studies in particular have drawn attention to the way in which natural knowledge became a site of contested authority in the early modern period: Eric H. Ash, *Power, Knowledge, and Expertise in Elizabethan England* (Baltimore: Johns Hopkins University Press, 2004); Antonio Barrera, "Local Herbs, Global Medicines: Commerce, Knowledge, and Commodities in Spanish America," in *Merchants and Marvels: Commerce, Science, and Art in Early Modern Europe*, ed. Pamela H. Smith and Paula Findlen (New York: Routledge, 2002), 163–81; Alison Sandman, "Mirroring the World: Sea Charts, Navigation, and Territorial Claims in Sixteenth-Century Spain," in Smith and Findlen, *Merchants and Marvels*, 83–108; and Deborah E. Harkness, "'Strange' Ideas and 'English' Knowledge: Natural Science Exchange in Elizabethan London," in Smith and Findlen, *Merchants and Marvels*, 138–62.

22. Stephen Greenblatt, *Renaissance Self-Fashioning: From More to Shakespeare* (Chicago: University of Chicago Press, 1980); Davis, *The Return of Martin Guerre*; Perez Zagorin, *Ways of Lying: Dissimulation, Persecution, and Conformity in Early Modern Europe* (Cambridge, Mass: Harvard University Press, 1990); Mark Crane, Richard Raiswell, and Margaret Reeves, eds., *Shell Games: Studies in Scams, Frauds, and Deceits (1300–1500)* (Toronto: Centre for Reformation and Renaissance Studies, 2004).

23. In *The Business of Alchemy*, Pamela Smith emphasized the way in which alchemy served as a kind of metaphorical bridge between traditional princely pursuits and a more modern, mercantilist economic policy. I am pushing this argument one step farther, suggesting that alchemical practices also had economic value in their own right. I will develop this argument in chap. 3.

24. In an important article, Lawrence Principe and William Newman suggest using the term "chymistry" in studies of the early modern period to avoid the false dichotomy between a premodern (and presumably prescientific) "alchemy" and a modern (and presumably scientific) "chemistry." While I wholeheartedly agree with their basic point, I prefer the term "alchemy" for this study because it

(or rather the German *Alchemie*) is the term that appears most frequently in the sources for this book. I use the term "alchemy" throughout this study, therefore, with the understanding that I do not intend it to indicate a set of practices somehow preceding, conflicting with, or unrelated to the "new science" associated with the Scientific Revolution. William R. Newman and Lawrence M. Principe, "Alchemy vs. Chemistry: The Etymological Origins of a Historiographic Mistake," *Early Science and Medicine* 3 (1998): 32–65.

25. "Bragadini Works On," 19 January 1590, in *The Fugger News-Letters, Being a Selection of Unpublished Letters from the Correspondents of the House of Fugger during the Years 1568–1605*, ed. Victor Klarwill, trans. Pauline de Chary (Bodley Head: John Lane, 1928), 158.

CHAPTER ONE

1. [*Miscellanea alchemica XXIV*], 1543, Wellcome Library, MS.524, fol. 1. This manuscript is a transcription of *Alchimia und Bergwerck* (Strasbourg: Jacob Cammerlander, 1534).

2. [*B.C.*], 1588, Wellcome Library, MS.107, fol. 1v.

3. [*Alchemy 16th cent.*], ca. 1525, Wellcome Library, MS.24, fol. 1. On the idea of secrets of nature, see William Eamon, *Science and the Secrets of Nature: Books of Secrets in Medieval and Early Modern Culture* (Princeton, N.J.: Princeton University Press, 1994).

4. Bruce Moran suggests that, "although a sprinkling of interest may be found in the subject within the university, it was, as a manual art, always denied a part in the scholastic curriculum." Bruce T. Moran, *Distilling Knowledge: Alchemy, Chemistry, and the Scientific Revolution* (Cambridge, Mass.: Harvard University Press, 2005), 34. William Newman, on the other hand, argues that alchemy's problem in the universities was that the *Sciant artifices*, assumed to have been written by Aristotle, seemed to condemn it. William R. Newman, *Promethean Ambitions: Alchemy and the Quest to Perfect Nature* (Chicago: University of Chicago Press, 2004), 72.

5. Michael Maier, *Examen fucorum pseudo-chymicorum* (Frankfurt: Theodor de Bry, 1617), 14; reprinted in Wolfgang Beck, "Michael Maiers Examen Fucorum Pseudo-Chymicorum: Eine Schrift wider den falschen Alchemisten" (PhD diss., Fakultät für Chemie, Biologie und Geowissenschaft der Technischen Universität München, 1992).

6. "Schreiben des Hofmeisters Abraham Bellin an den Kammersekretär Johann Sattler über ein Magd zu Frankfurt, Oder, die durch berichten Gold und Silber gemacht haben soll," 1598–99, HStA Stuttgart, Bestand A47, Bü. 3, Nr. 7.

7. Philipp Sömmering to Duke Julius of Braunschweig-Wolfenbüttel, 25 August 1573, NStA Wolfenbüttel, 1 Alt 9, Nr. 306, fol. 76.

8. Ibid.

9. The alchemical texts attributed to Villanova in the fourteenth century are most likely not authentic; Arnald himself found alchemists "ignorant" and "fool-

ish." See Michela Pereira, "Teorie dell'elixir nell'alchemia medievale," *Micrologus* 3 (1995): 103–48; William R. Newman, *The Summa Perfectionis of Pseudo-Geber: A Critical Edition, Translation and Study* (Leiden: E. J. Brill, 1991), 194; Charles Coulston Gillispie, ed., *Dictionary of Scientific Biography* (New York: Scribner, 1970–80), s.v. "Arnald of Villanova"; and Claus Priesner and Karin Figala, eds., *Alchemie: Lexikon einer hermetischen Wissenschaft* (Munich: Beck, 1998), s.v. "Arnald von Villanova."

10. Philipp Sömmering to Duke Julius of Braunschweig-Wolfenbüttel, 25 August 1573, NStA Wolfenbüttel, 1 Alt 9, Nr. 306, fol. 77. Aquinas repeatedly argued that art in general, and alchemical arts in particular, were weaker than nature, thus rejecting the alchemists' claims that they could make true metals artificially; William Newman has attributed Aquinas's rejection of alchemy to his "conservatism with regard to the art-nature distinction." See Newman, *Promethean Ambitions,* 51, 94–97. Despite this dim view of alchemy, however, many spurious works were attributed to Aquinas in the fifteenth century. Roger Bacon discussed alchemy in his *Opus minus, Opus tertium,* and *Epistola de secretis operibus artis et naturae* (all written before 1270). On Bacon, see Priesner and Figala, *Alchemie,* s.v. "Roger Bacon," and the bibliography there. Although he clearly was valued as an important alchemical authority in the sixteenth century and was frequently cited, very little is known about Bernhard of Treviso (also known as Bernardus or Bernhardus Trevisanus, Bernard the Trevisan, Bernardus Trevirensis, Bernard of Trevisan, and Bernhard of Trevigo). He rejected the sulfur-mercury theory in his famous letter to Christine de Pisan's father, Thomas of Bologna, in favor of the idea that mercury alone was the basis of all metals. This letter eventually was published as "Bernardi Trevirensis ad Thomam de Bononia medicum Regis Caroli octavi Responsio," in Bernhardus Trevisanus, *Morieni Romani. Item, primum in Lucem prodit Bernardi Trevirensis Responsio ad Thomam* (Paris: Gulielmum Guillard, 1564), 35–66; however, this text certainly circulated in manuscript much earlier. Several additional treatises were later ascribed to him (probably spuriously), including the autobiography in Bernard of Trevisan et al., *De chymico miraculo, quod lapidem philosophiae appellant* (Basel: Ex officina haeredum Petri Pernae, 1583), as well as manuscript copies and translations of "Bernard's process"; see, for example, "Processus Lapide Philosophici particularis et universalis . . . cum practica Bernhardii Comitis Taruisini maxime conueniens, Anno 1566," in Wellcome Library, MS.518, fols. 22–46; despite the Latin title, the text is in German. On Bernard, see Gillispie, *Dictionary of Scientific Biography,* s.v. "Bernard of Trevisan"; and Priesner and Figala, *Alchemie,* s.v. "Bernardus Trevirensis."

11. Sömmering described these texts only as "scolia in Bernhardum, testamentum Hermetis, aristochum Joannis [illegible—perhaps Trithemius?]" and a "kleines Pergamentbüchlein." He may have been referring to the *Testamentum* (1332), the oldest and most influential text of the pseudo-Lullian corpus. See William R. Newman, "The Philosopher's Egg: Theory and Practice in the Alchemy of Roger Bacon," *Micrologus* 3 (1995): 75–101; and Michela Pereira, *"Mater*

Medicinarum: English Physicians and the Alchemical Elixir in the Fifteenth Century," in *Medicine from the Black Death to the French Disease*, ed. Roger French (Aldershot: Ashgate, 1998), 26–52, and "Alchemy and the Use of Vernacular Languages in the Late Middle Ages," *Speculum* 74, no. 2 (1999): 336–56. See also Lynn Thorndike, *A History of Magic and Experimental Science* (New York: Columbia University Press, 1934–58), 4:3–64. Sömmering urged Julius to read the *scolia in Bernhardum* "with diligence . . . because the treasure of the secret is to be discovered in it, and, apart from the process, it is the dearest little book I have." Philipp Sömmering to Duke Julius of Braunschweig-Wolfenbüttel, 25 August 1573, NStA Wolfenbüttel, 1 Alt 9, Nr. 306, fol. 77. I am grateful to William Newman for his assistance in identifying these texts.

12. Johann Sternhals's 1488 treatise was printed in 1595 and again in 1680: Johann Sternhals, *Ritter Krieg; Das ist, ein philosophisch Gedicht, in Form eines gerichtlichen Process . . .* ([Erffordt]: Martin Wittel, 1595), and *Ritter-Krieg, Das ist, ein philosophisch-Geschicht* (Hamburg: Georg Wolff, 1680).

13. For a study of the role of ancient texts in the practice of Renaissance astrologer Girolamo Cardano, see Anthony Grafton, *Cardano's Cosmos: The Worlds and Works of a Renaissance Astrologer* (Cambridge, Mass.: Harvard University Press, 1999).

14. Alchemy's earliest texts, written in Greek in the first through the third centuries CE, were inherited and taken up by natural philosophers writing in Arabic in the seventh through the tenth centuries.

15. The word alchemy itself (*al-kimia*) came from Arabic, as did alkali (*al-qali*), alcohol (*al-kuhl*), and many others. For a good list of chemical terminology derived from Arabic, see E. J. Holmyard, *Alchemy*, 2nd ed. (Baltimore: Penguin Books, 1968), 10–11.

16. In 1144 Robert of Chester began his translation of an Arabic alchemical treatise, the "Book of the Composition of Alchemy," by commenting on alchemy's novelty in Europe: "Since what Alchymia is, and what its composition is, your Latin world does not yet know, I will explain in the present book." Quoted and translated in ibid., 106.

17. William Newman, "Technology and the Alchemical Debate in the Late Middle Ages," *Isis* 80 (1989): 430–37. See also chap. 2 of his *Promethean Ambitions*.

18. On medieval Latin alchemy, see Holmyard, *Alchemy*, 105–52. See also Bernhard Dietrich Haage, *Alchemie im Mittelalter: Ideen und Bilder von Zosimos bis Paracelsus* (Zurich: Artemis & Winkler, 1996); and Newman, "Technology and the Alchemical Debate in the Late Middle Ages," and *The Summa Perfectionis of Pseudo-Geber*.

19. Pereira, "Alchemy and the Use of Vernacular Languages," 339.

20. Udo Benzenhöfer, *Johannes de Rupescissa, "Liber de consideratione quintae essentiae omnium rerum" deutsch: Studien zur Alchemia medica des 15. bis 17. Jahrhunderts mit kritischer Edition des Textes* (Stuttgart: Steiner Verlag, 1989). On Rupescissa, see Leah DeVun, "John of Rupescissa and the States of Nature:

Science, Apocalypse, and Society in the Late Middle Ages" (PhD diss., Columbia University, 2004).

21. Herwig Buntz, "Deutsche alchimistische Traktate des 15. und 16. Jahrhunderts" (PhD diss., University of Munich, 1969).

22. Pereira, "Alchemy and the Use of Vernacular Languages," 346–47, 356. Pereira's dating of *Sol und Luna* differs from Joachim Telle's somewhat later dating. See Joachim Telle, *Sol und Luna: Literatur- und alchemiegeschichtliche Studien zu einem altdeutschen Bildgedicht* (Hürtgenwald: Guido Pressler Verlag, 1980).

23. Joachim Telle, "Bemerkungen zum 'Rosarium Philosophorum,'" in *Rosarium Philosophorum: Ein Alchemisches Florilegium des Spätmittelalters, Faksimile der illustrierten Erstausgabe Frankfurt 1550,* ed. Joachim Telle, trans. Lutz Claren and Joachim Huber, 2 vols. (Weinheim: VCH Verlagsgesellschaft, 1992).

24. See, for example, Bernard of Trevisan et al., *De chymico miraculo,* and *Opuscula chemica* (Leipzig: Johanes Rosen, 1605); and Ramón Lull, *De alchimia opuscula quae sequuntur* (Nuremberg: Johan Petreius, 1546), *Libelli aliquot chemici* (Basel: Peter Perna, 1572), and *Raymundii Lullii opera* (Strasbourg: Lazar Zetzner, 1598).

25. Rudolf Hirsch, "The Invention of Printing and the Diffusion of Alchemical and Chemical Knowledge," *Chymia* 3 (1950): 115–42.

26. Frances A. Yates, *Giordano Bruno and the Hermetic Tradition* (Chicago: University of Chicago Press, 1964).

27. See, for example, Peter Forshaw, "'Alchemy in the Amphitheatre': Some Considerations of the Alchemical Content of the Engravings in Heinrich Khunrath's *Amphitheatre of Eternal Wisdom* (1609)," in *Art and Alchemy,* ed. Jacob Wamberg (Copenhagen: Museum Tusculanum Press, 2006), 195–220.

28. On commonplace books, see Ann Blair, "Humanist Methods in Natural Philosophy: The Commonplace Book," *Journal of the History of Ideas* 53 (1992): 541–51; Ann Moss, *Printed Commonplace-Books and the Structuring of Renaissance Thought* (Oxford: Clarendon Press, 1996); and Yates, *Giordano Bruno and the Hermetic Tradition,* 163.

29. *De Alchimia* (Nuremberg: Johannes Petreius, 1541).

30. See also the *Theatrum chemicum,* 6 vols. (Strasbourg: Lazarus Zetzner, 1602–61). On early printing and alchemy, see Hirsch, "The Invention of Printing"; Telle, "Bemerkungen zum 'Rosarium Philosophorum,'" 163–64; and Lynn Thorndike, "Alchemy during the First Half of the Sixteenth Century," *Ambix* 2 (1938): 26–37.

31. See Rodovský's collected translations in the library of the National Museum in Prague: "Bavora ml. Rodovského z Hustiřan, sborník alchymistických překladů," 1578, Narodní Muzeum Praha, IV C 20 (372). On Rodovský generally, see Otakar Zachar, *O alchymii a českých alchymistech* (Prague: V. Kotrba, 1911), and "Z dějin alchymie v Čechách. I. Bavor mladši Rodovský z Hustiřan, alchymista český," *Časopis Musea Kralovstí Českého* XXI–II (1899–1900): 157–63.

32. Rodovský tried to gain financial support to translate Paracelsus's works

into Czech in 1573. Bavor Rodovský to Vilém Rožmberk, 6 February 1573, SOA Třeboň RRA 25. The entire text of this letter is reprinted in Zachar, *O alchymii a českých alchymistech,* 175ff.

33. Bavor mladši Rodovský z Hustiřan to Vilém Rožmberk, 6 February 1573, SOA Třeboň RRA 25.

34. There is a voluminous literature on Paracelsus and his followers. Standard works include Allen G. Debus, *The Chemical Philosophy: Paracelsian Science and Medicine in the Sixteenth and Seventeenth Centuries* (New York: Science History Publications, 1977); Walter Pagel, *Paracelsus : An Introduction to Philosophical Medicine in the Era of the Renaissance* (Basel: Karger, 1982); and Karl Sudhoff, *Paracelsus: Ein deutsches Lebensbild aus den Tagen der Renaissance* (Leipzig: Bibliographisches Institut, 1936). See also Heinz Dopsch, Kurt Goldammer, and Peter F. Kramml, *Paracelsus (1493–1541): Keines andern Knecht* (Salzburg: Anton Pustet, 1993); Heinz Schott and Ilana Zinguer, eds., *Paracelsus und seine Internationale Rezeption in der Frühen Neuzeit: Beiträge zur Geschichte des Paracelsismus,* Brill's Studies in Intellectual History, vol. 86 (Leiden: Brill, 1998); Joachim Telle, ed., *Analecta Paracelsica: Studiena zum Nachleben Theophrast von Hohenheims im deutschen Kulturgebiet der Frühen Neuzeit,* Heidelberger Studien zur Naturkunde der frühen Neuzeit, vol. 4 (Stuttgart: F. Steiner, 1994); and Joachim Telle, ed., *Parerga Paracelsica: Paracelsus in Vergangenheit und Gegenwart,* Heidelberger Studien zur Naturkunde der frühen Neuzeit, vol. 3 (Stuttgart: Franz Steiner Verlag, 1991).

35. See Priesner and Figala, *Alchemie,* s.v. "Dorn, Gerhard"; and Didier Kahn, "Les débuts de Gerhard Dorn," in Telle, *Analecta Paracelsica,* 59–126.

36. Petrus Severinus, *Idea medicinae philosophicae, fundamenta continens totius doctrinae Paracelsicae, Hippocraticae, et Galenicae* (Basel: Sixtus Henricpetrus, 1571). On Severinus, see Jole Shackelford, *A Philosophical Path for Paracelsian Medicine: The Ideas, Intellectual Context, and Influence of Petrus Severinus, 1540–1602* (Copenhagen: Museum Tusculanum, 2004).

37. Gerhard Dorn and Adam von Bodenstein, *Dictionarium Theophrasti Paracelsi* (Frankfort: Christoff Rab, 1584; reprint, Hildesheim: G. Olms, 1981); Adam von Bodenstein, *Onomasticon Theophrasti Paracelsi* (Basel: Perna, 1575); Leonhard Thurneisser zum Thurm, *Melitsah Kai Hermâeneia = das ist, Onomasticum und Interpretatio, oder aussführliche Erklerung* (Berlin: Nicolas Voltzen, 1574–83); Michael Toxites, *Onomastica II* ([Strasbourg: Per Bernhardum Jobinum], 1574; reprint, Hildesheim: G. Olms, 1984).

38. Johannes Huser, *Erster[-zehender] Theil der Bücher und Schriften . . . jetzt auffs neu . . . an Tag geben . . . ,* 10 vols. (Basel: Conrad Waldkirch, 1589–90).

39. Paracelsus, *Paragranum,* vol. 8 of *Sämtliche Werke,* ed. Karl Sudhoff (Munich: R. Oldenbourg, 1922–23), 186, as quoted in Wolf-Dieter Müller-Jahncke and Julian Paulus, "Die Stellung des Paracelsus in der Alchemie," in Dopsch, Goldammer, and Kramml, *Paracelsus,* 149–50.

40. Paracelsus, *De mineralibus,* vol. 3 of *Sämtliche Werke,* 32, as quoted in

Joachim Telle, "'Vom Stein der Weisen': Ein alchemoparacelsstische Lehrdichtung des 16. Jahrhunderts," in Telle, *Analecta Paracelsica,* 167–212.

41. Telle, "'Vom Stein der Weisen,'" 177. Paracelsians who did understand Paraclesus's disapproval of transmutational alchemy, moreover, often simply ignored his efforts to distance alchemy from goldmaking. The Paracelsian physician and mining expert Leonhard Thurneisser (1531–96), for example, otherwise never discussed transmutational alchemy in his work but wrote a single treatise in 1584 which described in detail the transmutation of copper into gold. This text, *De tranmutatione veneris in solem,* has puzzled historians who see a contradiction in a Paracelsian's pursuit of alchemical gold. However, it might also suggest how little conflict many practitioners actually saw between simultaneously pursuing Paracelsian medicine and alchemical gold. On this treatise and its place in Thurneisser's work, see Peter Morys, "Leonhard Thurneissers *De transmutatione veneris in solem,*" in *Die Alchemie in der europäischen Kultur- und Wissenschaftsgeschichte,* ed. Christoph Meinel (Wiesbaden: Otto Harrassowitz, 1986), 85–98. Even Johannes Huser, the Paracelsian physician who published and edited Paracelsus's collected works in 1589–90, did not entirely heed the admonition to pursue alchemy solely for its medical, not metallurgical, virtues. Joachim Telle has noted that Huser's efforts to obtain a treatise by Georg Klett (1467–1513) on the philosophers' stone "evinces none of the dull antiquarian or philological-editorial tendencies of the bookish scholar, but rather a thoroughly praxis-oriented technical interest in alchemia transmutatoria." Telle also notes, "In complete contrast to Hohenheim's quest to 'pharmaceuticize' *alchemia transmutatoria,* in the mental world of his editor Huser aspects of traditional alchemy—[such as] the 'work,' or rather 'goldmaking,' a cementation method directed at the production of gold, or the practices of the fraudulent goldmaker Marco Bragadino (ca. 1545/50–1591)—play a central role." Joachim Telle, "Johann Huser und der Paracelsismus im 16. Jahrhundert," in Dopsch, Goldammer, and Kramml, *Paracelsus,* 348.

42. Telle, "Johann Huser," 342; and Müller-Jahncke and Paulus, "Die Stellung des Paracelsus in der Alchemie," 151.

43. Telle noted, "'Vom Stein der Weisen' [On the philosophers' stone] generally is among the German didactic poems of alchemical content passed down most often." Telle, "'Vom Stein der Weisen,'" 181–82. The treatise was first attributed to Paracelsus in "De lapide," a treatise written in 1575 by David Beuther, an alchemist working for Elector August of Saxony. See Telle, "'Vom Stein der Weisen,'" 193.

44. Müller-Jahncke and Paulus, "Die Stellung des Paracelsus in der Alchemie," 151.

45. Although, as Eamon points out, these *Kunstbücher* largely have been ignored by historians of alchemy. See Eamon, *Science and the Secrets of Nature,* 112–33.

46. For the somewhat confusing publishing history of this *Kunstbuch,* see Eamon, *Science and the Secrets of Nature,* 114–16; and Ernst Darmstaedter, *Berg-,*

Probir- und Kunstbüchlein, Münchner Beiträge zur Geschichte und Literatur der Naturwissenschaften und Medizin, vols. 2 and 3 (Munich: Verlag der Münchner Drucke, 1926).

47. [Peter Kertzenmacher], *Alchimia, Das ist alle Farben, Wasser, Olea, Salia, und Alumina . . . zubereyten* (Frankfurt am Main: Bey C. Engenoffs Erben, 1570), 9. For more on these particular substances and their use in alchemy, see Priesner and Figala, *Alchemie*, 378, 166, 370, 159, 313.

48. Kertzenmacher, *Alchimia*, 40ff.

49. Kertzenmacher, *Alchimia, 3*.

50. Ann Blair, "Reading Strategies for Coping with Information Overload, ca. 1550–1700." *Journal of the History of Ideas* 64, no. 1 (2003): 11–28.

51. Philipp Sömmering to Duke Julius of Braunschweig-Wolfenbüttel, 25 August 1573, NStA Wolfenbüttel, 1 Alt 9, Nr. 306, fols. 76–76v. I appreciate Andrew Sparling's suggestions for this translation.

52. Pereira, "Alchemy and the Use of Vernacular Languages," 353. On alchemical reading practices in early modern England, see Lauren Kassell, "Reading for the Philosophers' Stone," in *Books and the Sciences in History*, ed. Marina Frasca-Spada and Nick Jardine (Cambridge: Cambridge University Press, 2000), 132–50.

53. On this idea in antiquity, see Eamon, *Science and the Secrets of Nature*, chap. 1.

54. For an overview of the alchemical *donum dei*, see Vladimír Karpenko, "Alchemy as 'Donum Dei,'" *Hyle: An International Journal for the Philosophy of Chemistry* 4 (1998): 63–80.

55. Ibid.

56. Cristoff von Hirschenberg in Prag to Vilém Rožmberk, 25 October 1583, SOA Třeboň RRA 25.

57. Sternhals, *Ritter Krieg*, preface.

58. Thomas Norton, *The Ordinall of Alchimy, in Theatrum chemicum britannicum containing severall poeticall pieces of our famous English philosophers, who have written the hermetique mysteries in their owne ancient language*, ed. Elias Ashmole (London: Printed by J. Grismond for Nath. Brooke, 1652; reprint, Baltimore: Williams and Wilkins Co., 1031), 14.

59. Ibid., 33.

60. Ibid., 33–34.

61. Ibid., 13–14.

62. Norton, *Ordinall*, as quoted in John Read, *From Alchemy to Chemistry* (Toronto: Dover Publications, 1931), 85.

63. "Was Philipp Sömmering anfangs und zum Eingang in den gute berichtet und ausgesagt," n.d. [ca. 1574], NStA Wolfenbüttel, 1 Alt 9, Nr. 311, fol 14ff.

64. Paracelsus, *Die grosse Wundarznei*, vol. 10 of *Sämtliche Werke*, 20, as quoted in Udo Benzenhöfer, *Paracelsus* (Reinbek bei Hamburg: Rowohlt Taschenbuch Verlag, 1997), 34.

65. Christof von Hirschenberg to Vilém Rožmberk, 25 October 1583, SOA Třeboň RRA 25.

66. "Vertrag zwischen Herzog Johann Friedrich von Württemberg und seine Mutter Sibille, geb. Fürstin zu Anhalt, einerseits, und den 3 Freunden Andreas Reich, Johann Andreas Hess u. Sebastian Hesch anderseits," 27 March 1610, HStA Stuttgart, Bü. 5, Nr. 6, fols. 1v–2.

67. Ibid., fol. 2. On Reiche, see Hans-Georg Hofacker, *". . . sonderlich hohe Künste und vortreffliche Geheimnis": Alchemie am Hof Herzog Friedrichs I. von Württemberg—1593 bis* 1608 (Stuttgart: Verein der Freunde des Chemischen Instituts Dr. Flad e. V., 1993), 30–31.

68. "Almighty God / From great Doctours hath this *Science* forbod, / And granted it to few Men of his mercy, / Such as be faithfull trew and lowly." Norton, *Ordinall,* as quoted in Read, *From Alchemy to Chemistry,* 85.

69. Norton, *Ordinall,* 3.

70. Vannoccio Biringuccio, *La pirotechnia* (Venice: Venturino Roffinello, Ad instatntia di Curio Nau. and Fratelli, 1540), reprinted as *The Pirotechnia of Vannoccio Biringuccio: The Classic Sixteenth-Century Treatise on Ores and Assaying,* ed. and trans. Cyril Stanley Smith and Martha Teach Gnudi, rev. ed. (1959; facsimile, New York: Dover Publications, 1990). Quote on p. 335 of reprint edition.

71. Other examples abound. For instance, when the Prague burgher Petr Hlavsa left his position as mint master for the kingdom of Bohemia in 1561, he began working as the manager of the Bohemian magnate Vilém Rožmberk's Prague alchemical operations. Hlavsa's alleged introduction of a new smelting technique in the Kutná Hora silver mines in Bohemia may have driven the royal chamber into debt and caused his downfall. Josef Svatek, *Culturhistorische Bilder aus Böhmen* (Vienna: Wilhelm Braunmüller, k. k. Hof- und Universitätsbuchhändler, 1879), 62.

72. On Thurneisser, see Paul H. Boerlin, *Leonhard Thurneysser als Auftraggeber* (Basel: Birkhäuser Verlag, 1970); Morys, "Leonhard Thurneissers *De transmutatione veneris in solem,*" and *Medizin und Pharmazie in der Kosmologie Leonhard Thurneissers zum Thurm (1531–1596),* Abhandlungen zur Geschichte der Medizin und der Naturwissenschaften, Heft 43, ed. Rolf Winau and Heinz Müller-Dietz (Husum: Matthiesen, 1982); and Gabriele Spitzer, *". . . und die Spree führt Gold": Leonhard Thurneysser zum Thurn, Astrologe—Alchimist—Arzt und Drucker im Berlin des 16. Jahrhunderts* (Wiesbaden: Reichert, 1996).

73. Maier, *Examen fucorum pseudo-chymicorum,* 17.

74. Moran, *Distilling Knowledge,* 11–25; Rudolf Werner Soukup and Helmut Mayer, *Alchemistisches Gold, Paracelsistische Pharmaka,* ed. Helmuth Grössing, Karl Kadletz, and Marianne Klemun, Perspektiven der Wissenschaftsgeschichte, vol. 10 (Vienna: Böhlau Verlag, 1997), 120–72.

75. Recall also that one of the first operations Sömmering learned was the art of distillation.

76. "Verhör Georg Harts in München," 22 December 1595, HStA Stuttgart, Bestand A47, Bü. 1, Nr. 2.

77. To the question "whether he didn't [present] himself as an artisan [*Künstler*] and [claim] that he could make gold out of silver," Hört "said absolutely not, because David [Pirkheimer] took him with him to Stuttgart for the sole reason that he knows how to deal with the fire and with distilling, and he would be a good *Laborant.*" "Verhör Georg Harts in München," 22 December 1595, HStA Stuttgart, Bestand A47, Bü. 1, Nr. 2.

78. The Bohemian historian Josef Svatek, in fact, referred to two of these regions (the Erzgebirge and Joachimstal) as "alchemical academies." Svatek, *Culturhistorische Bilder aus Böhmen,* 63.

79. Or so, at least, believed the chief superintendent of mines for the Holy Roman Emperor, Lazar Ercker. See Lazarus Ercker, *Treatise on Ores and Assaying,* trans. and ed. Anneliese Grünhaldt Sisco and Cyril Stanley Smith (Chicago: University of Chicago Press, 1951).

80. Maier, *Examen fucorum pseudo-chymicorum,* 17.

81. *Coelum philosophorum; seu, Secreta naturae* (Paris: Vivantius Gaultherot, 1543).

82. Heinrich Wolff et al., *Herliche medicische [sic] Tractat, vor nie in Truck kommen* (Strasbourg: Bernhart Jobin, 1576).

83. For an overview of this literature, see Moran, *Distilling Knowledge,* 11–25; Telle, "Bemerkungen Zum 'Rosarium Philosophorum,'" 163.

84. Pamela O. Long, "The Openness of Knowledge: An Ideal and Its Context in 16th-Century Writings on Mining and Metallurgy," *Technology and Culture* 32, no. 2 (1991): 320.

85. Ibid., 328–29.

86. On these two early mining pamphlets, see ibid., 328–30; and Darmstaedter, *Berg-, Probir- und Kunstbüchlein.* A third important treatise was Biringuccio, *La pirotechnia.* The fact that it appeared only in Italian in the sixteenth century, however, would have limited its central European audience.

87. The *Bergbüchlein,* for example, discussed the sulfur-mercury theory of the composition of metals, which Arabic alchemists introduced into alchemy.

88. Georgius Agricola, *De re metallica* (German vernacular edition) (Basel: Froben, 1561), and *De re metallica libri XII* (Basel: Froben, 1556).

89. Georg Agricola, *De re metallica,* trans. and ed. Herbert Clark Hoover and Lou Henry Hoover (New York: Dover, 1950), 248; Ercker, *Treatise on Ores and Assaying,* 9.

90. Long, "The Openness of Knowledge," 323.

91. Ibid., 353.

92. Ercker, *Treatise on Ores and Assaying,* 19–34.

93. On distilling mercury ore, see ibid., 285–86. On making and using *aqua fortis,* or "parting acid," see ibid., 131–61. Recipes for *aqua fortis* also appeared in the Italian Benvenuto Cellini's treatise on goldsmithing and sculpture, Benvenuto

Cellini, *The Treatises of Benvenuto Cellini on Goldsmithing and Sculpture,* trans. C. R. Ashbee (Bow, England: Laurence Hodson, 1898; reprint, New York: Dover Books, 1967), 105ff.; as well as Kertzenmacher's *Alchimia,* 20–21; and Biringuccio's mining treatise, *The Pirotechnia of Vannoccio Biringuccio,* ed. and trans. Smith and Gnudi, 183–88, 91–96.

94. Michael Maier, *Atalanta Fugiens* (Oppenheim: Theodor de Bry, 1618), preface. I have relied here on the English translation by Joscelyn Godwin in Michael Maier, *Atalanta Fugiens: An Edition of the Fugues, Emblems, and Epigrams* (Grand Rapids, MI: Phanes Press, 1989), 101. See also Maier, *Examen fucorum pseudo-chymicorum,* esp. 13–20.

95. Maier, *Examen fucorum pseudo-chymicorum,* 16.

96. Ibid., 16–17.

97. Ibid., 16–17.

98. Agricola, *De re metallica,* trans. and ed. Hoover and Hoover, 3–4.

99. For example, the "Treuhertzige Warnungs-Vermahnung eines Getreuen Liebhabers der Warheit an alle wahre Liebhaber der Naturgemässen Alchymiae Transmutatoriae; daß wegen der Bübischen Handgriffe der betriegerischen Arg-Chymisten gute Auffacht vonnöthen," which was appended to Heinrich Khunrath's *Von hylealischen, das ist, pri-materialischen Catholischen oder allgemeinen natürlichen Chaos: Der naturgemässen Alchymiae und Alchymisten* (Magdeburg, 1597; reprint, Graz: Akademische Druck- u. Verlagsanstalt, 1990).

CHAPTER TWO

1. René van Bastelaer, *The Prints of Peter Brueghel the Elder: Catalogue Raissoné,* new ed. (San Francisco: Alan Wofsy Fine Arts, 1992), 263–67, original quote and translation on 64. The "enigmatic" Latin inscription, which was not included with Brueghel's original drawing for this engraving, may be translated as follows: "The ignorant ought to put up with things and afterwards labor diligently. The juice of the precious Stone, common but then rare, is a certain single thing, vile but found everywhere, mingled with the four natures, crammed in a cloud, no mineral thing, and while of the first rank is such that it is found everywhere at hand." Lawrence M. Principe and Lloyd DeWitt, *Transmutations: Alchemy in Art* (Philadelphia: Chemical Heritage Foundation, 2002), 11.

2. Art historians have offered a variety of interpretations of this image, which was engraved by Philips Galle and published by Hieronymus Cock in Antwerp. Peter Dreyer, for instance, has interpreted Brueghel's drawing as a representation of seduction into sin at the hands of the Antichrist, followed by redemption through God's grace at the poorhouse. Peter Dreyer, "Brueghels Alchimist von 1558: Versuch einer Deutung *ad sensum mysticum,*" *Jahrbuch der Berliner Museen (Jahrbuch der Preußischen Kunstsammlungen, Neue Folge)* 19 (1977): esp. 74ff. Jacques Van Lennep, on the other hand, has argued that the image is a sympathetic attempt to differentiate between a "real" philosophical alchemy aimed at a higher

truth and a "false" alchemy overly focused on the earthly desire for wealth. Jacques Van Lennep, "An Alchemical Message in Brueghel's Prints?" in *The Prints of Pieter Brueghel the Elder,* ed. David Freedberg (Tokyo: Tokyo Shibum, 1989). Most satisfactorily, to my mind, Philippe and Françoise Roberts-Jones have highlighted the many contradictory messages in Brueghel's print, rather than limiting Brueghel to a single interpretation: "This apparently unambiguous work contains contradictory ideas—each of which may find echoes, suggestions, even firm statements, to support it among the customs, literature, or art of the time." Philippe Roberts-Jones and Françoise Roberts-Jones, *Pieter Bruegel* (New York: Harry N. Abrams, 2002), 37.

3. Lorraine Daston and H. Otto Sibum, "Introduction: Scientific Personae and Their Histories," *Science in Context* 16, no. 1/2 (2003): 2, 3, 7. The rest of the essays in this volume collectively offer examples of how the concept of scientific persona can be used in the history of science. Mauss's original essay was first published as "Une catégorie de l'esprit humain: La notion de personne, celle de 'moi': Un plan de travail," *Journal of the Royal Anthropological Institute* 68 (1938): 263–81.

4. Daston and Sibum, "Introduction: Scientific Personae," 5.

5. Gadi Algazi, "Scholars in Households: Refiguring the Learned Habitus, 1480–1550," *Science in Context* 16, no. 1/2 (2003): 9–42.

6. William Newman has identified the author as Paul of Taranto. William R. Newman, ed., *The Summa Perfectionis of Pseudo-Geber: A Critical Edition, Translation and Study* (Leiden: E. J. Brill, 1991), 634. I have taken all quotes here from Newman's translation and critical edition of the *Summa perfectionis.*

7. Ibid., 635–38.

8. Ibid., 638–40.

9. Ibid., 638.

10. On the persona, or habitus, of the medieval scholar, see Algazi, "Scholars in Households," esp. 9–12.

11. Indeed, Bonus included Moses, David, Solomon, and John the Evangelist among the founders of alchemy. Chiara Crisciani, "The Conception of Alchemy as Expressed in the *Pretiosa margarita novella* of Petrus Bonus of Ferrara," *Ambix* 20, no. 3 (1973): 173 and 71 n. 26.

12. *Pretiosa margarita novella,* as cited in ibid., 170 n. 21.

13. *Pretiosa margarita novella,* 50b, as quoted in ibid., 172 n. 28. For more on Petrus Bonus and his role in the art-nature debate, see also William R. Newman, *Promethean Ambitions: Alchemy and the Quest to Perfect Nature* (Chicago: University of Chicago Press, 2004), 83–89.

14. Newman, *Promethean Ambitions,* 89–97.

15. On this topic, see Pamela H. Smith, *The Body of the Artisan: Art and Experience in the Scientific Revolution* (Chicago: University of Chicago Press, 2004).

16. Algazi, "Scholars in Households," 12. See also his "Food for Thought: Hieronymus Wolf Grapples with the Scholarly Habitus," in *Egodocuments and History: Autobiographical Writing in Its Social Context since the Middle Ages,*

ed. Rudolf Dekker (Hilversum: Verloren, 2002), 21–44, and "Gelehrte Zerstreutheit und gelernet Vergeßlichkeit: Bemerkungen zu ihrer Rolle in der Formierung des Gelehrtenhabitus," in *Der Fehltritt: Vergehen und Versehen un der Vormoderne*, ed. Peter von Moos (Cologne: Böhlau, 2001), 235–50.

17. Pamela O. Long, *Openness, Secrecy, Authorship: Technical Arts and the Culture of Knowledge from Antiquity to the Renaissance* (Baltimore: Johns Hopkins University Press, 2001); Smith, *The Body of the Artisan*.

18. Anna Zieglerin, for instance, certainly drew on this dimension of alchemical tradition in the 1570s when she claimed that her alchemical knowledge, received through a kind of divine revelation, would prepare the world for the Last Days. See Tara E. Nummedal, "Alchemical Reproduction and the Career of Anna Maria Zieglerin," *Ambix* 48 (2001): 56–68.

19. Newman, *Promethean Ambitions*; Smith, *The Body of the Artisan*.

20. On sixteenth-century satirical tradition, see Barbara Könneker, *Satire im 16. Jahrhundert: Epoche, Werke, Wirkung* (Munich: C. H. Beck, 1991).

21. For overviews of literary representations of the alchemist in medieval Europe, see W. Ganzenmüller, *Die Alchemie im Mittelalter* (Paderborn, 1938; reprint, Hildesheim, 1967), 90–96; Will H. L. Ogrinc, "Western Society and Alchemy from 1200–1500," *Journal of Medieval History* 6 (1980): 103–37, esp. 8–14; Michela Pereira, "Alchemy and the Use of Vernacular Languages in the Late Middle Ages," *Speculum* 74, no. 2 (1999): 336–56; and Jost Weyer, "Der Alchemist im lateinischen Mittelalter (13. bis 15. Jahrhundert)," in *Der Chemiker im Wandel der Zeiten: Skizzen zur geschichtlichen Entwicklung des Berufbildes*, ed. Eberhard Schmauderer and Gesellschaft Deutscher Chemiker, Fachgruppe Geschichte der Chemie (Weinheim: Verlag Chemie, 1973), 11–41. Stanton J. Linden, *Darke Hierogliphicks: Alchemy in English Literature from Chaucer to the Restoration* (Lexington: University Press of Kentucky, 1996), offers a useful overview of medieval and early modern English literary representations of alchemy.

22. Newman, *Promethean Ambitions*, esp. chap. 2.

23. Ogrinc, "Western Society and Alchemy," 124; Pereira, "Alchemy and the Use of Vernacular Languages."

24. A few literary authors did engage more substantively with the scholastic debate about art and nature. As Newman has pointed out, Jean de Meun's elaboration on the *Roman de la Rose* in the 1270s "transferred the dry and technical arguments of writers such as pseudo–Roger Bacon and Paul of Taranto directly into the context of courtly love, where they would be evaluated by an audience comprising all levels of literate society." Newman, *Promethean Ambitions*, 82. Dante encounters two alchemists among the falsifiers in the tenth pit of the eighth and lowest circle of hell in canto 29 of the *Inferno*.

25. For an introduction to late medieval and early modern English literary representations of alchemy, see Linden, *Darke Hierogliphicks*.

26. Francesco Petrarca, *Von der Artzney bayder Glück, des guten und widerwertigen* (Augsburg: Heinrich Stayner, 1532). For a modern facsimile edition, see

Von der Artzney bayder Glueck, des guten und widerwertigen, ed. Manfred Lemmer (Hamburg: Friedrich Wittig Verlag, 1984).

27. Indeed, Brant was directly involved in producing the extraordinarily successful illustrated German edition of 1532, as the publisher's preface noted; he also contributed a poetic preface to the 1532 edition. Sebastian Brant, *Das Narrenschyf* (Basel: Johann Bergmann von Olpe, 1494), and *Stultifera navis,* trans. Jacob Locher (Basel: Johann Bergmann de Olpe, 1497). This jurist, Latin poet, municipal secretary, professor, and author, however, was also deeply involved in editing for various publishing houses in Basel and elsewhere. Edwin Zeydel, for instance, estimates that "at least one-third of all the volumes printed in Basel before 1490 show signs of his collaboration, whether they mention his name or not." Edwin H. Zeydel, ed. and trans., *The Ship of Fools, by Sebatian Brant* (New York: Columbia University Press, 1944), 4. Regarding Brant's important role in the 1532 Augsburg translation of *De remediis,* Walther Scheidig noted: "The poet and philosopher Petrarca was the author, the German humanist and educator Sebastian Brant the editor." Scheidig, *Die Holzschnitte des Petrarca-Meisters: Zu Petrarcas Werk von der Artzney bayder Glück des guten und widerwärtigen, Augsburg 1532* (Berlin: Henschelverlag/Deutsche Akademie der Künste, 1955), 8. Desiderius Erasmus, *Familiarum colloquiorum* (Basel: Johannes Froben, 1524).

28. Heinrich Cornelius Agrippa von Nettesheim, *De incercitudine et vanitate scientiarum atque artium declamatio* (Antwerp: Grapheus Drucker, 1530); Johann Clajus, *Altkumistica, das ist: Die Kunst, aus Mist durch seine Wirckung, Gold zu machen: Wider die betrieglichen Alchimisten, und ungeschickte vermeinte Theophrasisten* (Leipzig: Z. Bärwaldt, 1586).

29. The 1532 German edition translates the conversation partners as *Vernunfft* [reason] and *Freud* [joy] conversing, whereas the Latin has Reason and Hope (*Spes*) conversing. See Conrad H. Rawski, ed., *Remedies for Fortune Fair and Foul: A Modern English Translation of De remediis utriusque fortune, with a Commentary* (Bloomington: Indiana University Press, 1991), 299–301.

30. I have relied heavily on the English translation in Rawski, *Remedies for Fortune Fair and Foul,* 1:299–301. However, I have made occasional alterations in order to better reflect the sense of the 1532 German edition.

31. Ibid., 1:300.

32. Ibid., 1:301.

33. Scheidig, *Die Holzschnitte des Petrarca-Meisters.* The identity of the "Master of Petrarch," as the woodcut designer has come to be known, was lost in the century after he first designed the woodcuts in 1519. While some have identified him as Hans Weiditz (before 1500–ca. 1536), the designer of the woodcuts for Otto Brunfels's well-known herbal from the same period, others have disputed this attribution. For a concise overview of the dispute about the identity of the Master of Petrarch, see Kristin Lohse Belkin, "Weiditz: (2) Hans Weiditz (ii)," *Grove Art Online* http://groveart.com/shared/views/article.html?section=art.090980.3 (accessed 14 November 2005), and the bibliography cited there.

34. Rawski, *Remedies for Fortune Fair and Foul,* 1:300.

35. "Heynrich Steyner Drucker zu Augsburg, wünscht allen Lesern dises büchs, hayl, und die genad Gottes," in Petrarca, *Von der Artzney bayder Glück.*

36. The *Narrenschiff* appeared in six authorized editions and at least six pirated editions before his death in 1512; the original German text was also translated into Latin (*Stultifera navis,* 1497), which edition then became the basis for subsequent translations into French, English, and Dutch, making possible the book's international fame. See Eckhard Bernstein, *German Humanism* (Boston: Twayne Publishers, 1983), 44. On the additions to each of the various editions, see Zeydel, *The Ship of Fools,* 21–31. For an overview of Brant's life and work, as well as the *Narrenschiff,* see Bernstein, *German Humanism,* 39–48. On the satirical tradition, see also Könneker, *Satire im 16. Jahrhundert,* esp. 54–64.

37. John Walter Van Cleve, *The Problem of Wealth in the Literature of Luther's Germany* (Columbia, S.C.: Camden House, 1991), 55.

38. As Zeydel explains in his annotation to this passage, "The word is *guckuss.* There may be a pun involved. The cuckoo might refer to the foolishness of the alchemist, to the fact that he, like the cuckoo, lays his eggs in others' nests to be hatched, and the word *gucken,* 'to look' or 'to peep,' may also be in Brant's mind." Zeydel, *The Ship of Fools,* 388 n. 6.

39. Zeydel, *The Ship of Fools,* 329.

40. The humanist, alchemist, and natural magician Heinrich Cornelius Agrippa von Nettesheim repeated many of these themes in his *De incercitudine et vanitate scientiarum atque artium declamatio* (written in 1526). Agrippa's relationship to alchemy was much more complicated than this text reveals, however. See Christopher I. Lehrich, *The Language of Demons and Angels: Cornelius Agrippa's Occult Philosophy* (Leiden: Brill, 2003), 76–93.

41. Desiderius Erasmus, "Desiderius Erasmus Roterodamus to the Reader," in *De utilitate colloquiorum* (Basel: Johannes Froben, 1526), as cited in *The Colloquies of Erasmus,* trans. Craig R. Thompson (Chicago: University of Chicago Press, 1965), 631.

42. Erasmus, "Altcumistica," in *The Colloquies of Erasmus,* 239–44.

43. Ibid., 244–45.

44. Geoffrey Chaucer, *The Canterbury Tales: A Verse Translation,* trans. David Wright (Oxford: Oxford University Press), 435.

45. Rawski, *Remedies for Fortune Fair and Foul,* 1:300–301.

46. For a modern German reprint of the *Narrenschiff* see Sebastian Brant, *Das Narrenschiff,* ed. Manfred Lemmer (Stuttgart: Reclam, 1966). The chapter "Falsch und Beschiß" appears on 174–77. For an English verse rendering of Brant's original passage on alchemy, see Zeydel, *The Ship of Fools,* 328–30. Although Zeydel's translation is more poetic, I have used my own translation to reflect more literally Brant's original German.

47. Van Cleve, *The Problem of Wealth,* 52.

48. Brant, *Das Narrenschiff,* ed. Lemmer, 177.

49. It appeared, for example, in the "Canon's Yeoman's Tale" in Chaucer's *Canterbury Tales.*

50. Brant, *Das Narrenschiff,* ed. Lemmer, 176.

51. The identities of the woodcut designers for the *Narrenschiff* are uncertain, as is the exact nature of Brant's involvement in their production. In 1892, Daniel Burckhardt ascribed some of the woodcuts to Dürer, who spent time in Basel from 1492 to 1494 and met Brant there; this possibility continues to appear in the scholarly literature. Other scholars, however, notably Heinrich Wölfflin, have disputed this attribution. Daniel Burckhardt, *Albrecht Dürer's Aufenthalt in Basel 1492–1494* (Munich: G. Hirt, 1892); Heinrich Wölfflin, *Handzeichnungen* (Munich: R. Piper & Co., 1919). For an overview of the woodcuts and dispute about their creators, see Zeydel, *The Ship of Fools,* 19–21; and Friedrich Winkler, *Dürer und die Illustrationen zum Narrenschiff: Die Baseler und Straßburger Arbeiten des Künstlers und der altdeutsche Holzschnitt* (Berlin: Deutscher Verein für Kunstwissenschaft, 1951). On the relationship between word and image in *Von der Artzney bayder Glück,* see Hans-Joachim Raupp, "Die Illustrationen zu Francesco Petrarca, 'Von der Artzney Bayder Glueck des Guten und Widerwertigen' (Augsburg 1532)," *Wallraf-Richartz-Jahrbuch (Westdeutsches Jahrbuch für Kunstgeschichte)* 45 (1984): 59–112.

52. Thus argued Peter Dreyer in "Brueghels Alchimist," 74–90.

53. On Brant's views of wealth in the *Narrenschiff,* see Van Cleve, *The Problem of Wealth,* 23–61.

54. *The Colloquies of Erasmus,* 248–54. Interestingly, Robert Boyle included a character named Philoponus, "Lover of Work," to defend the chymists in his *Dialogue on Transmutation.* One wonders whether Boyle had Erasmus in mind when he named this character. See Lawrence M. Principe, *The Aspiring Adept: Robert Boyle and His Alchemical Quest* (Princeton, N.J.: Princeton University Press, 1998), 98–103.

55. See, for instance, the *Liber vagatorum,* on which see Robert Jütte, *Abbild und soziale Wirklichkeit des Bettler- und Gaunertums zu Beginn der Neuzeit: Sozial-, mentalitäts- und sprachgeschichtliche Studien zum Liber vagatorum (1510)* (Cologne: Böhlau, 1988). For a brief overview of rogues' literature as a "realistic" reflection of marginalized society, see Jütte, *Poverty and Deviance in Early Modern Europe* (Cambridge: Cambridge University Press, 1994), 178–85.

56. On the criminalization of poverty in sixteenth-century Europe, see Jütte, *Poverty and Deviance,* chaps. 8 and 9; and Pieter Spierenburg, "The Body and the State: Early Modern Europe," in *The Oxford History of the Prison: The Practice of Punishment in Western Society,* ed. Norval Morris and David J. Rothman (New York: Oxford University Press, 1995). For the Beham image, see Jütte, *Poverty and Deviance,* 16.

57. Agrippa von Nettesheim, *De incertitudine et vanitate scientiarum atque artium declamatio,* 264.

58. Whether or not the rogues' literature was reflective of the real conditions of

life for vagabonds and beggars is a contested issue. I will return to this point at the end of this chapter, below.

59. Agrippa von Nettesheim, *De incertitudine et vanitate scientiarum atque artium declamatio,* 266.

60. Clajus, *Altkumistica.*

61. Grimm's *Wörterbuch* defines *Finantzerey* as "fraudatio, betrug, wucher." Jacob Grimm and Wilhelm Grimm, *Deutsches Wörterbuch* (Leipzig: S. Hirzel, 1854). Also available at http://www.woerterbuchnetz.de/woerterbuecher/dwb/wbgui?lemid=GF04326 (accessed 13 December 2005).

62. Clajus, *Altkumistica.*

63. For a parallel example of the way in which fraud could serve as a vehicle for the performance of expertise in natural history, see Paula Findlen, "Inventing Nature: Commerce, Art, and Science in the Early Modern Cabinet of Curiosities," in *Merchants and Marvels: Commerce, Science, and Art in Early Modern Europe,* ed. Pamela H. Smith and Paula Findlen (New York: Routledge, 2002), 297–323.

64. In addition to the texts discussed here, see, for instance, the "Lectori salutem" in Andreas Libavius, *Alchemia Andreae Libavii* (Frankfurt am Main: Excudebat J. Saurius impensis Petri Kopffii, 1597); the "Vorrede" to Franciscus Epimentheus [Hieronymus Reusner], *Pandora, Das ist, Die Edleste Gab Gottes* (Basel: [S. Apario], 1582); Alexander Lauterwald, *Colloquium Philosophicum* (Cologne: Heinrich Netessem, 1597), and *Widerlegung der Altkuhmisterey* (Cologne: Nettesheim, 1597); Ewald Hoghelande, *Kurtzer Bericht und klarer Beweiss,* trans. Joachim Tancke (Leipzig: M. Lantzenberger for J. Apel, 1604); and Theobald Hoghelande, *Von den Irrwegen der Alchemisten* (Frankfurt am Main: Johann Spiess, 1600); and "Colloquium Chymica," in *Cabala chymica,* by Franz Kieser, Alexander von Suchten, and Georgius Clettus (Mülhausen: bey Martin Spiessen: in Verlegung Johann Spiessen &c., 1606), 283ff.

65. Theophilus Caesar, ed., *Alchimyspiegel: oder Kurtz entworffene Practick, der gantzen Chimischen Kunst* (Frankfurt am Main: Vincentii Steinmeyer, 1597).

66. Ibid., "Vorrede."

67. Guglielmo Gratarolo, ed., *Verae alchemiae artisque metallicae* (Basel: Henricus Petri & Petrus Perna, 1561), and *Alchemiae, quam vocant, artisque metallicae* (Basel: Henricus Petri & Petrus Perna, 1572).

68. Caesar, *Alchimyspiegel,* 40.

69. Ibid., 42.

70. Ibid., 47.

71. Ibid., 54–56.

72. Ibid., 42, 44, 56.

73. Eric H. Ash, *Power, Knowledge, and Expertise in Elizabethan England* (Baltimore: Johns Hopkins University Press, 2004).

74. Leonhard Thurneisser, *Magna alchymia* (Berlin: Niclaus Voltz, 1583), "Vorrede" (unpaginated).

75. Ibid.

76. Ibid.

77. Thurneisser's puns here are untranslatable. He describes the false gems as "schöne *glasiasticos* von Schnarchgacken, Schafsphiren, Diebßmanten, Spillnellen, und was der Steinen mehr sind." Ibid.

78. Ibid.

79. Ibid.

80. Ibid.

81. Thrasybulus is a Greek form of Khun-rath, "Bold in Counsel," and Ricenus is anagram of Khunrath's first name, Heinrich, or, in Latin, Enricus. I am grateful to Peter Forshaw for his assistance with the authorship of this text.

82. Heinrich Khunrath, *Von hylealischen, das ist pri-materialischen Catholischen oder allgemeinen natürlichen Chaos: Der naturgemässen Alchymiae und Alchmysten* (Magdeburg, 1597; reprint, Graz, Akademische Druck- u. Verlagsanstalt, 1990), "Vorrede und Apologie deß Auctoris," 7.

83. For example, in *Von hylealischen . . . Chaos*, 259–63, as well as the *Alchemical Citidel* and *Calumniators* engravings in his *Amphitheatrum sapientiae aeternae* (Hanau: G. Antonius, 1609). See Peter Forshaw, "'Alchemy in the Amphitheatre': Some Considerations of the Alchemical Content of the Engravings in Heinrich Khunrath's *Amphitheatre of Eternal Wisdom* (1609)," in *Art and Alchemy*, ed. Jacob Wamberg (Copenhagen: Museum Tusculanum Press, 2006), 195–220.

84. Thrasybulus Ricenus [Heinrich Khunrath], "Treuhertzige Warnungs-Vermahnung eines Getreuen Liebhabers der Warheit an alle wahre Liebhaber der Naturgemässen Alchymiae Transmutatoriae; daß wegen der Bübischen Handgriffe der betriegerischen Arg-Chymisten gute Auffacht vonnöthen," in Khunrath, *Von hylealischen . . . Chaos*, 268. Michael Maier would later quote this text extensively, attributing it to Khunrath, in his *Examen fucorum pseudo-chymicorum* (Frankfurt am Main: Theodor de Bry, 1617); reprinted in Wolfgang Beck, "Michael Maiers Examen Fucorum Pseudo-Chymicorum: Eine Schrift wider den falschen Alchemisten" (PhD diss., Fakultät für Chemie, Biologie und Geowissenschaft der Technischen Universität München, 1992). See also David Beuther, *Davids Beuthers . . . universal, und vollkommener Bericht* (Frankfurt: In Verlegung Wilhelm Fitzers, 1631), esp. 46–47.

85. Ricenus, "Treuhertzige Warnungs-Vermahnung," 283.

86. Ibid., 286.

87. Ibid., 282.

88. This was a common argument. See, for instance, Lauterwald, *Widerlegung der Altkuhmisterey*.

89. Andreas Libavius offered a similar justification in the preface to his *Alchemia*, where he argued that his text would empower people to judge for themselves whether a particular alchemist was an impostor or not. Libavius, *Alchemia*, "Lectori salutem."

90. William Eamon, *Science and the Secrets of Nature: Books of Secrets in*

Medieval and Early Modern Culture (Princeton, N.J.: Princeton University Press, 1994), esp. chap. 4.

91. For the latter view, see, for example, Linden, *Darke Hierogliphicks,* esp. chap. 1; Petra Schramm, *Die Alchemisten: Gelehrte—Goldmacher—Gaukler. Ein dokumentarischer Bildband* (Wiesbaden: Edition Rarissima, 1984); and Weyer, "Der Alchemist im lateinischen Mittelalter (13. bis 15. Jahrhundert)," esp. 14–25. One important exception is Sylvain Matton, "L'influence de l'humanisme sur la tradition alchimique," *Micrologus* 3 (1995): 279–345.

92. The connection I have drawn between the rogue's literature and *Betrüger* literature of the sixteenth century provides a useful analogy. Although social, economic, and literary historians have long debated whether these representations of roguery can be used as evidence for the organization and culture of a subculture of vagabonds, beggars, and criminals, the best work now employs a sophisticated understanding of both literary and social history sources, as well as the way in which these interacted. For an overview of this problem, see Jütte, *Poverty and Deviance,* 178–85.

93. Daston and Sibum, "Introduction: Scientific Personae," 5.

CHAPTER THREE

1. For the older view, see Henry Carrington Bolton, *The Follies of Science at the Court of Rudolph II, 1576–1612* (Milwaukee: Pharmaceutical Review Publishing Co., 1904); Theodor Wagner, "Wissenschaftlicher Schwindel aus dem südlichen Böhmen," *Mittheilungen des Vereins für Geschichte der Deutschen in Böhmen* 16 (1878): 112–23; and R. J. W. Evans's historiographical discussion of this perspective in his *Rudolf II and His World: A Study in Intellectual History* (Oxford: Oxford University Press, 1973), 44–47. Evans's *Rudolf II and His World* offered an important revision of this view, as did Thomas DaCosta Kaufman, *The Mastery of Nature: Aspects of Art, Science, and Humanism in the Renaissance* (Princeton, N.J.: Princeton University Press, 1993); Bruce Moran, *The Alchemical World of the German Court: Occult Philosophy and Chemical Medicine in the Circle of Moritz of Hessen (1572–1632)* (Stuttgart: Franz Steiner Verlag, 1991), and "German Prince-Practitioners: Aspects in the Development of Courtly Science," *Technology and Culture* 22, no. 2 (1981): 253–74; Pamela H. Smith, "Alchemy as a Language of Mediation at the Habsburg Court," *Isis* 85 (March 1994): 1–25; and Erich Trunz, *Wissenschaft und Kunst im Kreise Kaiser Rudolfs II. 1576–1612* (Neumünster: Karl Wachholtz Verlag, 1992), 13.

2. Although he focuses largely on the philosophical and medical alchemy characteristic of the court of Moritz of Hessen, Bruce Moran suggestively points to other kinds of alchemical patronage in his conclusion, noting in particular the way in which "alchemical projects could easily be turned into types of court technology" at other German courts in this period. See Moran, *The Alchemical World of the German Court,* 171–76, and "German Prince-Practitioners."

3. "Dokumente über Conrad von Grumbach und Bischof Julius Echter von Mespelbrunn 1593," Staatsarchiv Würzburg, Histor. Saal VII, Sign. 25/380, as quoted in Jost Weyer, *Graf Wolfgang II. von Hohenlohe und die Alchemie: Alchemistische Studien in Schloß Weikersheim, 1587–1610,* ed. Historischen Verein für Württembergisch Franken, Stadtarchiv Schwäbisch Hall and Hohenlohe-Zentralarchiv Neuenstein, Forschungen aus Württembergisch Franken, vol. 39 (Sigmaringen: Jan Thorbecke Verlag, 1992), 311.

4. These negotiations were part of a much longer history between the prince-bishops of Würzburg and the Grumbach family. Conrad von Grumbach's father Wilhelm had long been in the service of the Würzburger prince-bishops as a Frankisch *Reichsritter.* In the 1560s, however, Wilhelm von Grumbach turned against Julius's predecessor as prince-bishop, Melchior von Zobel, and his Lutheran ally Elector August of Saxony. Von Zobel was killed in the military campaign that ensued. In 1567, Elector August of Saxony besieged Gotha, where Grumbach and the duke of Ernestine Saxony Johann Friedrich II were making a last stand, and eventually imprisoned Duke Johann Friedrich II and executed Grumbach. (Incidentally, it was during this siege that Sömmering fled Gotha, eventually winding up at Duke Julius's court in Wolfenbüttel.) The executed Wilhelm's son Conrad von Grumbach reconciled with the new Würzburger bishop, Julius, and in 1568 Conrad regained most of his father's lands, including the castle and village at Rimpar, which had been the Grumbach family's seat since the late fourteenth century. Conrad inherited huge debts from his father, however, and was forced to sell the castle and village at Rimpar back to the bishopric in 1593. It was at this point that he entered into negotiations with the prince-bishop Julius, trading an alchemical process for continued possession of his family's lands. The outcome of the alchemical work is unknown, as there are no subsequent archival documents. Conrad von Grumbach was still alive (and still impoverished) in 1599. Weyer, *Graf Wolfgang II. von Hohenlohe und die Alchemie,* 310–11. On Duke Johann Friedrich II, Wilhelm von Grumbach, and the events that led up to their capture in 1567, see Peter Elsel Starenko, "In Luther's Wake: Duke John Frederick II of Saxony, Angelic Prophecy, and the Gotha Rebellion of 1567" (PhD thesis, University of California, Berkeley, 2002).

5. As the economic historian Hermann Kellenbenz put it, the problem was a result of "the heavy debt of the princes, the nobility, and the clergy, who, with their emphatically feudal barter economy arrangements, did not yet meet the demands of an economic rhythm that was determined more and more by the urban Bürgertum." Hermann Kellenbenz, *Deutsche Wirtschaftsgeschichte,* 2 vols. (Munich: Beck, 1977), 145.

6. Richard Bonney, *The European Dynastic States, 1494–1660* (Oxford: Oxford University Press, 1991), 423.

7. Reiner Groß, *Geschichte Sachsens* (Leipzig: Edition Leipzig, 2001), 72.

8. Although princes do not typically come to mind in the context of sixteenth-century economic development, Hans-Joachim Kraschewski and others have shown that some princes took a proactive role in developing their territories'

economies. See Hans-Joachim Kraschewski, *Wirtschaftspolitik im deutschen Territorialstaat des 16. Jahrhunderts: Herzog Julius von Braunschweig-Wolfenbüttel*, ed. Ingomar Bog, Neue Wirtschaftsgeschichte, vol. 15 (Cologne and Vienna: Böhlau Verlag, 1978), particularly the historiographical discussion of this problem on 4–12.

9. Phillippe Braunstein, "Innovations in Mining and Metal Production in Europe in the Late Middle Ages," *Journal of European Economic History* 12, no. 3 (1983): 574.

10. Danuta Molenda, "Technological Innovation in Central Europe between the XIVth and the XVIIth Centuries," *Journal of European Economic History* 17, no. 1 (1988): 73–74.

11. For a technical description of the *Saigerprozess* in particular, see Braunstein, "Innovations," 587–88. John Munro attributes the spur for new technologies to "the closure of many mints and a prolonged deflation, which in turn enhanced the real value (purchasing power) of silver." John H. Munro, "Patterns of Trade, Money, and Credit," in *Handbook of European History, 1400–1600: Late Middle Ages, Renaissance, and Reformation*, ed. Thomas A. Brady, Heiko Augustinus Oberman, and James D. Tracy (Leiden: E. J. Brill, 1994), 167. These arguments are admittedly reductionist, assuming an inexhaustible desire for more profit as the driving force behind technological and financial innovation; the subject of late medieval technological innovation would benefit from a more critical rethinking in this sense.

12. On this new technology and associated costs, see Braunstein, "Innovations"; Kellenbenz, *Deutsche Wirtschaftsgeschichte*, 162–64; and William J. Wright, "The Nature of Early Capitalism," in *Germany: A New Social and Economic History*, ed. Bob Scribner (London: Arnold, 1996), 189–92.

13. The first printed treatise on mining, the *Bergbüchlein* (ca. 1505–10), offers a contemporary description of how these shares worked. See *Bergwerk- und Probierbüchlein: A Translation from the German of the "Bergbüchlein," a Sixteenth-Century Book on Mining Geology and of the "Probierbüchlein," a Sixteenth-Century Work on Assaying*, ed. Annelise Grünhaldt Sisco and Cyril Stanley Smith (New York: American Institute of Mining and Metallurgical Engineers, 1949), 18–19. This move to offer *Kuxen* had profound consequences for the miners, however. As Hermann Kellenbenz and numerous other historians have pointed out, this was the moment when "labour and capital separated" because miners were no longer working for themselves in collectives, but rather for outside investors. Hermann Kellenbenz, *The Rise of the European Economy: An Economic History of Continental Europe from the Fifteenth to the Eighteenth Century* (New York: Holmes and Meier Publishers, 1976), 80; Susan C. Karant-Nunn, "The Women of the Saxon Silver Mines," in *Women in Reformation and Counter-Reformation Europe*, ed. Sherrin Marshall (Bloomington: Indiana University Press, 1989), 24–46.

14. On the Fuggers' mining interests and relationships with political elites, see Jacob Strieder, *Jacob Fugger the Rich, Merchant and Banker of Augsburg,*

1459–1525 (New York: Adelphi Co., 1931), 105–66. Other German metal enterprises included the Welser, Paumgartner, Manlich, Haug, Link, and Langnauer, which are discussed in Kellenbenz, *The Rise of the European Economy*, 79–80.

15. On regalian rights, see W. Ebel, "Über das landesherrliche Bergregal," *Zeitschrift für Bergrecht* 109 (1968): 146–83; and Ekkehard Henschke, *Landesherrschaft und Bergbauwirtschaft: Zur Wirtschafts- und Verwaltungsgeschichte des Oberharzer Bergbaugebietes im 15. und 17. Jahrhundert*, Schriften zur Wirtschafts- und Sozialgeschichte, ed. Wolfram Fischer, vol. 23 (Berlin: Duncker & Humbolt, 1974), 30–41.

16. Braunstein, "Innovations," 581–82.

17. Wright, "The Nature of Early Capitalism," 192–93. For Duke Georg of Saxony's involvement in the tin trade, see Kellenbenz, *The Rise of the European Economy*, 105–6; and Jacob Strieder, *Studien zur Geschichte kapitalistischer Organisationsformen: Monopole, Kartelle und Aktiengesellschaften im Mittelalter und zu Beginn der Neuzeit*, 2nd expanded ed. (New York: B. Franklin, 1971), 212–20.

18. Henschke, *Landesherrschaft und Bergbauwirtschaft*, 42–43. For a telling example of why this kind of expertise was so important, see Eric H. Ash, "Queen v. Northumberland, and the Control of Technical Expertise," *History of Science* 39 (2001): 215–40, and, more broadly, Ash, *Power, Knowledge, and Expertise in Elizabethan England* (Baltimore: Johns Hopkins University Press, 2004).

19. Braunstein, "Innovations," 582.

20. Strieder, *Studien zur Geschichte kapitalistischer Organisationsformen*, 212–20; Wright, "The Nature of Early Capitalism," 192–94.

21. Braunstein, "Innovations"; Molenda, "Technological Innovation"; John H. Munro, "The Central European Silver Mining Boom, Mint Outputs, and Prices in the Low Countries and England. 1450–1550," in *Money, Coins, and Commerce: Essays in the Monetary History of Asia and Europe from Antiquity to Modern Times*, ed. Eddy Van Cauwenberghe (Leuven: Leuven University Press, 1991), 119–83.

22. Munro, "Patterns of Trade, Money, and Credit," 167.

23. A. Laube, *Studien über den erzgebirgischen Silberbergbau von 1470 bis 1546* (Berlin: Akademie-Verlag, 1974), 282. See also pp. 268–69 for the yields of individual mines in Saxony.

24. On the Fuggers' mining interests and relationships with political elites, see Strieder, *Jacob Fugger the Rich*, 105–66.

25. Munro, "Patterns of Trade, Money, and Credit," 172.

26. Karl Czok and Reiner Gross, "Das Kurfürstentum, die sächsisch-polnische Union und die Staatsreform (1547–1789)," in *Geschichte Sachsens*, ed. Karl Czok (Weimar: Hermann Böhlaus Nachfolger, 1989), 219.

27. Rudolf Werner Soukup and Helmut Mayer, *Alchemistisches Gold, Paracelsistische Pharmaka*, ed. Helmuth Grössing, Karl Kadletz, and Marianne Klemun, Perspektiven der Wissenschaftsgeschichte, vol. 10 (Vienna: Böhlau Verlag, 1997), 250.

28. Kellenbenz, *The Rise of the European Economy*, 114.

29. Munro, "Patterns of Trade, Money, and Credit," 172.

30. In Mexico Europeans began to mine silver in Zacatecas (1546), Compostele de la Guadalajara (1553), Sombrete and Durango (1555), Trestrillo (1562), and Parral (late sixteenth century). For an introduction to the technological, economic, and social dimensions of mining in the Americas, see the essays in P. J. Bakewell, *Mines of Silver and Gold in the Americas* (Brookfield, Vt.: Variorum, 1997). See also F. C. Spooner, "The Economy of Europe, 1559–1609," in *The New Cambridge Modern History,* vol. 3, ed. R. B. Wernham (Cambridge: Cambridge University Press, 1968), 25. On the role of silver in facilitating global trade, see Dennis O. Flynn and Arturo Giráldez, "Born with a 'Silver Spoon': The Origin of World Trade in 1571," *Journal of World History* 6, no. 2 (1995): 201–21.

31. Bonney, *The European Dynastic States,* 420.

32. Spooner, "The Economy of Europe," 26.

33. Ibid., 30.

34. Kellenbenz, *The Rise of the European Economy,* 32.

35. Ibid., 168 and 32; Spooner, "The Economy of Europe," 28.

36. Kellenbenz, *The Rise of the European Economy,* 32; Munro, "Patterns of Trade, Money, and Credit," 172–75.

37. On princely entrepreneurs, see in particular, Fritz Redlich, "Der deutsche fürstliche Unternehmer, eine typische Erscheinung des 16. Jahrhunderts," *Tradition: Zeitschrift für Firmengeschichte und Unternehmer-Biographie* 3, no. 1 (1958): 17–23, 98–112; and Wilhelm Treue, "Das Verhältnis von Fürst, Staat und Unternehmer in der Zeit des Merkantilismus," *Vierteljahrschrift für Sozial- und Wirtschaftsgeschichte* 44 (1957): 26–56.

38. The *regalia* afforded sovereigns the right "to operate mines themselves for whatever mineral they desired, to grant mining rights [to others, and] to enact statutory regulations governing mining procedures for every type of mine." Ebel, "Über das landesherrliche Bergregal," 182. For more components of the Bergregal, see Kraschewski, *Wirtschaftspolitik,* 54 n. 184.

39. The same year, Julius received 42 percent of his income from his mining enterprises in Rammelsberg and the Harz. Kraschewski, *Wirtschaftspolitik,* 152.

40. Ibid., 127–28. This text can also be found in Erasmus Ebener, "Bericht an Herzog Julius von Braunschweig vom 26. Januar 1572. Mit mineralogischen, metallurgischen und chemischen Anmerkungen vom Zehntner Meyer und vom Bergamts-Auditor Hausmann," *Hercynisches Archiv* (Halle) 1, no. 1–4 (1805): 494–540. Ebener (1511–77), was born in Nuremberg and studied in Wittenberg with Melanchthon, whom he accompanied to the Reichstag at Augsburg in 1530. He entered Julius's service in Wolfenbüttel in 1573 as councilor on, among other things, mining matters, and assisted him in founding the university in Helstedt. See Albert Rhamm, *Die betrüglichen Goldmacher am Hofe des Herzogs Julius von Braunschweig: Nach den Processakten* (Wolfenbüttel: Julius Zwißler, 1883), 90 n. 94.

41. Ebener, "Bericht an Herzog Julius," 494.

42. Kraschewski, *Wirtschaftspolitik,* 128.

43. Hans-Joachim Kraschewski, "Der 'ökonomische' Fürst: Herzog Julius als Unternehmer-Verleger der Wirtschaft seines Landes, besonders des Harz-Bergbaus," in *Staatsklugheit und Frömmigkeit: Herzog Julius von Braunschweig-Lüneburg—ein norddeutscher Landesherr des 16. Jahrhunderts,* ed. Christa Graefe and the Herzog August Bibliothek (Wolfenbüttel) (Weinheim: VCH Verlagsgesellschaft, Acta Humaniora, 1989), 45.

44. Antwort des Herzog Julius an Landgraf Wilhelm, 26. Januar 1576, Hessisches Staatsarchiv Marburg, 4 f Br.-W., 76, as quoted in Kraschewski, *Wirtschaftspolitik,* 126.

45. Kraschewski, *Wirtschaftspolitik,* 173–75. For a very informative study of the relationship between politics and mining in the Harz in the sixteenth and seventeenth centuries, see Henschke, *Landesherrschaft und Bergbauwirtschaft.*

46. Czok and Gross, "Das Kurfürstentum, die sächsisch-polnische Union und die Staatsreform (1547–1789)," 220 and 234.

47. Helen Watanabe-O'Kelly, *Court Culture in Dresden: From Renaissance to Baroque* (Houndmills: Palgrave, 2002), 78–88; Fritz Bönisch, *Kursächsische Kartographie bis zum Dreißigjährigen Krieg* (Berlin: Deutscher Verlag der Wissenschaften, 1990), 228–29.

48. Hugo Koch, *Sächsische Gartenkunst,* 2nd ed. (Beucha: Sax-Verlag, ca. 1999), 12–17.

49. Jutta Bäumel describes this book in her essay "Electoral Tools and Gardening Implements," in *Princely Splendor: The Dresden Court, 1580–1620,* ed. Dirk Syndram and Antje Scherner (Dresden: Staatliche Kunstsammlungen Dresden; Milan: Mondadori Electa, 2004). I have not been able to locate the 1571 edition; however, the catalog of the Herzog-August-Bibliothek in Wolfenbüttel lists a later edition: August, Elector of Saxony, *Künstlich Obst Garten Büchlein: Churfürst Augusti zu Sachsen . . . Jtzo auffs New von einem Liebhaber deß Gartenbawes wiederumb an Tag gegeben* (Berlin: Georg Runge; Martin Guth, printer, 1636). I am grateful to Alisha Rankin for pointing out this edition to me.

50. Moran, "German Prince-Practitioners." On Anna of Saxony in the wider context of German noblewomen and medicine, see Alisha Rankin, "Becoming an Expert Practitioner: Court Experimentalism and the Medical Skills of Anna of Saxony (1532–85)," *Isis* 98 (2007): 23–53, and "Medicine for the Uncommon Woman: Experience, Experiment, and Exchange in Early Modern Germany" (PhD diss., Harvard University, 2005).

51. Watanabe-O'Kelly, *Court Culture in Dresden,* 121–22.

52. Bäumel, "Electoral Tools and Gardening Implements," 162.

53. Groß, *Geschichte Sachsens,* 76–77; Rudolf Kötzschke and Hellmut Kretzschmar, *Sächsische Geschichte* (Frankfurt am Main: Verlag Wolfgang Weidlich, 1965), 221. On Stolpen and Ostra in particular, see Koch, *Sächsische Gartenkunst,* 12–13.

54. Czok and Gross, "Das Kurfürstentum, die sächsisch-polnische Union

und die Staatsreform (1547–1789)," 220 and 234; Kötzschke and Kretzschmar, *Sächsische Geschichte*, 221–22. On the Saxon economy in the first half of the sixteenth century, see Karlheinz Blaschke, *Sachsen im Zeitalter der Reformation* (Gütersloh: Gütersloher Verlagshaus, 1970), 33–47. See also the essays in Manfred Bachmann, Harald Marx, and Eberhard Wächtler, *Der Silberne Boden: Kunst und Bergbau in Sachsen* (Stuttgart: Deutsche Verlags-Anstalt; Leipzig: Edition Leipzig, 1990).

55. Hans-Georg Hofacker, *"... sonderlich hohe Künste und vortreffliche Geheimnis": Alchemie am Hof Herzog Friedrichs I. von Württemberg—1593 bis 1608.* (Stuttgart: Verein der Freunde des Chemischen Instituts Dr. Flad e. V., 1993), 9; David Sabean, *Power in the Blood: Popular Culture and Village Discourse in Early Modern Germany* (Cambridge: Cambridge University Press, 1984), 4; Paul Sauer, *Herzog Friedrich I. von Württemberg 1557–1608: Ungestümer Reformer und weltgewandter Autokrat* (Munich: Deutsche Verlags-Anstalt, 2003), 203–32.

56. Jean Bodin, *Respublica das ist: Gründtliche und rechte Underweysung, oder eigentlicher Bericht ... Jetzt ... auss lateinischer unnd frantzösischer Sprach, in unser ... Teutsch,* trans. Johann Oswaldt (Mumpelgart: Jacques Foillet, 1592).

57. Willi A. Boelcke, "Das Haus Württemberg und die wirtschaftliche Entwicklung des Landes," in *900 Jahre Haus Württemberg: Leben und Leistung für Land und Volk,* ed. Robert Uhland (Stuttgart: W. Kohlhammer, 1985), 644–52; Hofacker, *"... sonderlich hohe Künste,"* 11–12. On Duke Julius's projects on canals and roads, see Kraschewski, "Der 'ökonomische' Fürst," 41–43; and Evans, *Rudolf II and His World,* 2.

58. Pamela O. Long, "The Openness of Knowledge: An Ideal and Its Context in 16th-Century Writings on Mining and Metallurgy," *Technology and Culture* 32, no. 2 (1991): 325 and entire. For a fuller discussion of this issue, see her *Openness, Secrecy, Authorship: Technical Arts and the Culture of Knowledge from Antiquity to the Renaissance* (Baltimore: Johns Hopkins University Press, 2001), esp. chap. 6.

59. Long, "The Openness of Knowledge," 322.

60. Soukup and Mayer, *Alchemistisches Gold,* 251. On the relationship between alchemy and metallurgy in theory and practice, see also Robert Halleux, "L'alchemiste et l'essayeur," in *Die Alchemie in der europäischen Kultur- und Wissenschaftsgeschichte,* ed. Christoph Meinel (Wiesbaden: Harrassowitz, 1986), 277–92; and Lothar Suhling, "'Philosophisches' in der frühneuzeitlichen Berg- und Hüttenkunde: Metallogenese und Transmutation aus der Sicht montanischen Erfahrungswissens," in Meinel, *Die Alchemie,* 293–313.

61. "Neigung des Herzogs Julius zur Alchemie; Befehl an die Beamten," 5 June 1576, NStA Wolfenbüttel, 2 Alt 24, as quoted in Kraschewski, *Wirtschaftspolitik,* 159.

62. Kraschewski, *Wirtschaftspolitik,* 130–31. For more on Rhenanus, see Moran, *The Alchemical World of the German Court,* 76; Rhamm, *Die betrüglichen*

Goldmacher, 69 n. 9; M. Cramer, *M. Johannes Rhenanus, der Pfarrherr und Salzgräfe zu Allendorf* (Halle: Buchhandlung des Waisenhauses, 1879); and Johannes Rhenanus, *Das deutsche Salinenwesen im 16. Jahrhundert: Reiseberichte des Allendorfer Salzgrafen Johannes Rhenanus (um 1528–1589)*, ed. Hans-Henning Walter (Freiberg: Bergakademie Freiberg, 1989), 67–78. On salt mining in early modern central Europe, see Peter Piasecki, *Das deutsche Salinenwesen, 1550–1650: Invention—Innovation—Diffusion* (Idstein: Schulz-Kirchner Verlag, 1987).

63. Kraschewski, *Wirtschaftspolitik*, 130–31.

64. "Was Philip Sömmering anfangs und zum Eingang in den gute berichtet und ausgesagt," n.d. [July 1574], NStA Wolfenbüttel, 1 Alt 9, Nr. 311, fols. 15v–16. See also Rhamm, *Die betrüglichen Goldmacher*, 7–8.

65. "Was Philip Sömmering anfangs und zum Eingang in den gute berichtet und ausgesagt," n.d. [July 1574], NStA Wolfenbüttel, 1 Alt 9, Nr. 311, fols. 16r–v.

66. On the Gotha rebellion of 1567, see Starenko, "In Luther's Wake."

67. "Klagesartikel gegen Philipp, n.d. [before 5 July 1574]," NStA Wolfenbüttel, 1 Alt 9, Nr. 310, fol. 70.

68. As Pamela Long has pointed out, technical writing on mining and metallurgy after 1550 came to focus increasingly these subjects. See Long, "The Openness of Knowledge."

69. Five different methods were introduced earlier in the sixteenth century to deal with this, which may explain in part why the mine was bankrupt by the end of the century. Molenda, "Technological Innovation," 75. See also Václav Březan, *Životy Posledních Rožmberků*, ed. Jaroslav Pánek, 2 vols. (Prague: Svoboda Praha, 1985), 703.

70. On Uden: "Schreiben vom August 1576, Unterschrift und Datum fehlen," NStA Wolfenbüttel, 2 Alt 24, fols. 48r–56v. On Töpfer: "Schreiben des Theophil Töpfer an Sander vom 5.2.1575," Fach 2a, 10, Oberbergamt Clausthal.-Zellerfeld Archiv des Oberbergamtes, as quoted in Kraschewski, *Wirtschaftspolitik*, 159–60. Theophilactus Töpfer (or Döpfer) left a number of archival traces throughout the Holy Roman Empire. After submitting this proposal to Julius in 1575, he worked in Český Krumlov (Krumau) for Vilém Rožmberk in 1585 and then had a brief stint in Stuttgart in 1597. See Jakob Faber to Vilém Rožmberk, Krumlov, 5 February 1585, SOA Třeboň RRA 25; HStA Stuttgart, Bestand A47, Bü. 3, Nrs. 3 and 4. One can only imagine where he worked in the intervening years.

71. "Die angegeben Neue Alchimisten, Moritz Lam und Georg v. Minden betr., Vernehmungsprotokoll vom 6.6.1576," NStA Wolfenbüttel, 2 Alt 24, as quoted in Kraschewski, *Wirtschaftspolitik*, 158–59.

72. Ebener, "Bericht an Herzog Julius."

73. Georg Honauer to Emperor Rudolf II, 5 January 1597, HStA Stuttgart, Bestand A47, Bü. 1, Nr. 10.

74. Hans Müller, Hofgoldschmied, to Rudolf II, 13 February 1597, HStA Stuttgart, Bestand A47, Bü. 1, Nr. 10.

75. Georg Honauer to Emperor Rudolf II, 5 January 1597, HStA Stuttgart, Bestand 47, Bü. 1, Nr. 10.

76. The documents connected with this case are collected in "Anders Buch der angegebenen Künstler Schreiben und Berichte, 1580–91," SHStA Dresden, Geheimer Rat (Geheimes Archiv), Loc. 4418/3, fols. 354–78. For a comparable example, see the discussion of Lazar Ercker's proposal to Duke Julius in chap. 4, below. On Kramer's involvement in the lead and green vitriol trade, see Hans-Joachim Kraschewski, "Heinrich Cramer von Clausbruch und seine Handelsverbindungen mit Herzog Julius von Braunschweig-Wolfenbüttel: Zur Geschichte des Fernhandels mit Blei und Vitriol in der zweiten Hälfte des 16. Jahrhunderts," *Braunschweigisches Jahrbuch* 66 (1985): 115–28.

77. "1. Heinrich Kramer zu Leipzig," 4 October 1583, SHStA Dresden, Geheimer Rat (Geheimes Archiv), Loc. 4418/3, fol. 358.

78. Ibid., fol. 359.

79. The use of the term "inventor" had long been standard in contracts concerning funding for the construction of mining equipment. See Molenda, "Technological Innovation," 78–79.

80. For recent work on the history of intellectual property, see Mario Biagioli and Peter Galison, eds., *Scientific Authorship: Credit and Intellectual Property in Science* (New York: Routledge, 2002).

81. "1. Heinrich Kramer zu Leipzig," 4 October 1583, fol. 359r–v.

82. Ibid., fol. 358.

83. "2. Handlung mit Heinrichen von Klausbruck sonst Kramer gnandt, der Neuen Schmeltzkunst halbenn," 1583, SHStA Dresden, Geheimer Rat (Geheimes Archiv), Loc. 4418/3, fols. 361–62.

84. "1. Heinrich Kramer zu Leipzig," fol. 358v. See also "2. Handlung mit Heinrichen von Klausbruck sonst Kramer gnandt, der Neuen Schmeltzkunst halbenn," 1583, fol. 361v.

85. "3. Verlass mit Heinrich Kramern," 14 October 1583, fol. 364.

86. Elector August to Heinrich Kramer, 18 October 1583, SHStA Dresden, Geheimer Rat (Geheimes Archiv), Loc. 4418/3, fol. 374.

87. Pamela H. Smith, *The Business of Alchemy: Science and Culture in the Holy Roman Empire* (Princeton, N.J.: Princeton University Press, 1994), "Alchemy as a Language of Mediation," "Consumption and Credit: The Place of Alchemy in Johann Joachim Becher's Political Economy," in *Alchemy Revisited*, ed. Z. R. W. M. von Martels (Leiden: Brill, 1990), 215–21, and "Curing the Body Politic: Chemistry and Commerce at Court, 1664–70," in *Patronage and Institutions: Science, Technology, and Medicine at the European Court, 1500–1750*, ed. Bruce T. Moran (Rochester, N.Y.: Boydell Press, 1991), 195–209.

88. Smith, "Alchemy as a Language of Mediation," 3.

89. Ibid., 5.

90. Ibid., 25.

CHAPTER FOUR

1. Hans Heinrich Nüschler to Duke Friedrich of Württemberg, 15 July 1601, HStA Stuttgart, Bestand A47, Bü. 3, Nr. 10.

2. Ibid.

3. Ibid.

4. On the early modern patronage of science and medicine, see Mario Biagioli, *Galileo Courtier: The Practice of Science in the Culture of Absolutism* (Chicago: University of Chicago Press, 1994), and "Galileo's System of Patronage," *History of Science* 28 (1990): 1–62; Paula Findlen, *Possessing Nature: Museums, Collecting, and Scientific Culture in Early Modern Italy* (Berkeley: University of California Press, 1994), esp. chap. 8; Bruce T. Moran, ed., *Patronage and Institutions: Science, Technology, and Medicine at the European Court, 1500–1750* (Rochester, N.Y.: Boydell Press, 1991); Vivian Nutton, *Medicine at the Courts of Europe, 1500–1837*, Wellcome Institute Series in the History of Medicine (London: Routledge, 1990); and Stephen Pumphrey and Frances Dawbarn, "Science and Patronage in England, 1570–1625," *History of Science* 42, no. 2 (2004): 137–87.

5. On these "prince-practitioners," see Bruce Moran, *The Alchemical World of the German Court: Occult Philosophy and Chemical Medicine in the Circle of Moritz of Hessen (1572–1632)* (Stuttgart: Franz Steiner Verlag, 1991); Rudolf Werner Soukup and Helmut Mayer, *Alchemistisches Gold, Paracelsistische Pharmaka*, ed. Helmuth Grössing, Karl Kadletz, and Marianne Klemun, Perspektiven der Wissenschaftsgeschichte, vol. 10 (Vienna: Böhlau Verlag, 1997); and Jost Weyer, *Graf Wolfgang II. von Hohenlohe und die Alchemie: Alchemistische Studien in Schloß Weikersheim, 1587–1610*, ed. Historischen Verein für Württembergisch Franken, Stadtarchiv Schwäbisch Hall and Hohenlohe-Zentralarchiv Neuenstein, Forschungen aus Württembergisch Franken, vol. 39 (Sigmaringen: Jan Thorbecke Verlag, 1992).

6. Findlen, *Possessing Nature*, 365.

7. On patronage and particularly the role of brokers, see Biagioli, *Galileo Courtier*, 20–25; and Findlen, *Possessing Nature*, chap. 8.

8. Albert Rhamm, *Die betrüglichen Goldmacher am Hofe des Herzogs Julius von Braunschweig: Nach den Processakten* (Wolfenbüttel: Julius Zwißler, 1883), 8–9. Similarly, Salome Scheinpflugerin, the manager of Vilém Rožmberk's alchemical laboratory in Třeboň, urged Rožmberk to reconsider his initial refusal to grant Johannes Placotomus an audience: "Thus, my special request falls to Your Grace, that you would graciously accord the aforementioned Placotomus an audience." Salome Scheinpflugerin to Vilém Rožmberk, St. Bartholomew's Day, 1588. SOA Třeboň RRA 25.

9. Hans Härpfell to Wilhelm von Rosenberg, n.d. [after 1570], SOA Třeboň RRA 25. Wilhelm Haißhammer [or also Geishammer] worked for Duke Wilhelm V of Bavaria from 1572 to 1574. See his letters to Wilhelm V and his *Kammersecre-*

tario Stephan Schleich, fol. 46, 65, 55–57, 93–96b, 102, 186, 282, 327, Fürstensachen 426A, BHStA Munich.

10. On the role of trust and social capital in the construction of knowledge, see Steven Shapin, *A Social History of Truth: Civility and Science in Seventeenth-Century England,* Science and Its Conceptual Foundations (Chicago: University of Chicago Press, 1994).

11. On competition among alchemists at court, see Moran, *The Alchemical World of the German Court,* 145–62.

12. "Anders Buch der angegebenen Künstler Schreiben und Berichte, 1580–91," SHStA Dresden, Geheimer Rat (Gehimes Archiv), Loc. 4418/3. Among the other arts listed were *Sahm- und Ackerkunst, Pulverkunst, Salzkunst, Berckkunste, Ziegelbrennen, Muhlenkunst, Hebekunst, Schmelzkunst, Wasserkunst, Alaun, Salz, Kriegskunst, Geometria, Atronomia,* and *Silberbrennen.* These entries are all listed, in alphebetical order according to the name of the person who proposed them, in the *Registratur* to this volume.

13. "Pithoporius d. Christoff ubergiebet eine Kurtze erlerung was warheit, und Sophisterey sey, in Theologia, Asttronomia, und Medicina." Ibid.

14. On these two different kinds of transmutation, see Lawrence M. Principe, *The Aspiring Adept: Robert Boyle and His Alchemical Quest* (Princeton, N.J.: Princeton University Press, 1998), 76–80.

15. Christof von Hirschenberg to Vilém Rožmberk, 25 October 1583, SOA Třeboň RRA 25. On Hirschenberg, see Joachim Telle, "Der Alchemist im Rosengarten: Ein Gedicht von Christoph von Hirschenberg für Landgraf Wilhelm IV. von Hessen-Kassel und Graf Wilhlem von Zimmern," *Euphorion* 71 (1977): 283–305.

16. This strategy also resembles those of the "expert mediators" whom Eric H. Ash has described in sixteenth-century England in *Power, Knowledge, and Expertise in Elizabethan England* (Baltimore: Johns Hopkins University Press, 2004). On von Stuckhard, see Moran, *The Alchemical World of the German Court,* 145–47. For the contract that resulted from Hirschenberg's letter, see Hlawsa von Liboslav to Vilém Rožmberk, 18 January 1574, SOA Třeboň RRA 25.

17. Clearly Lazar Ercker did not identify himself as an alchemist; in fact, as Pamela Long correctly has pointed out, he was quite critical of alchemy in his published treatises. In practice, however, Ercker's projects seem little different from those of people like Hirschenberg and Polhaimer. Compare his *Prozeß* here, for example, to Polhaimer's below. On Ercker and alchemy, see in particular Pamela O. Long, "The Openness of Knowledge: An Ideal and Its Context in 16th-Century Writings on Mining and Metallurgy," *Technology and Culture* 32, no. 2 (1991): 318–55.

18. On Ercker in general, see Pamela O. Long, *Openness, Secrecy, Authorship: Technical Arts and the Culture of Knowledge from Antiquity to the Renaissance* (Baltimore: Johns Hopkins University Press, 2001), 188–91; and Ludmila Kubátová, Hans Prescher, and Werner Weisbach, *Lazarus Ercker (1528/30–1594): Probierer,*

Berg- und Münzmeister in Sachsen, Braunschweig und Böhmen (Leipzig: Deutscher Verlag für Grundstoffindustrie, 1994).

19. Lazar Ercker in Prague to Herzog Julius in Wolfenbüttel, 3 May 1585, NStA Wolfenbüttel, 1 Alt 9, Nr. 394, fols. 1–2.

20. Ibid., fol. 1.

21. Ibid., fol. 1.

22. Ibid., fol. 2.

23. On Hájek, see R. J. W. Evans, *Rudolf II and His World: A Study in Intellectual History* (Oxford: Oxford University Press, 1973), 203–04. On Mosanus, see Moran, *The Alchemical World of the German Court*, 79–86. Finally, on Hlavsa (or Hlawsa) and Scheinpflugerin's activities, see the letters in SOA Třeboň RRA 25.

24. On the figure of the expert in sixteenth-century England, see Ash, *Power, Knowledge, and Expertise.*

25. Friedrich ordered these small assays on 19 August, 16 September, and 10 October 1596, Nr. 7, Bü. 1, HStA Stuttgart.

26. For example, Marco Bragadino's. See Ivo Striedinger, *Der Goldmacher Marco Bragadino* (Munich: Theodor Ackerman, 1928).

27. Duke Friderich to Georg Gadner, 19 August 1596, HStA Stuttgart, Bü. 1, Nr. 7. The ducat weighed 3.49 g, with a fineness of 986‰ gold. Helmut Kahnt and Bernd Knorr, *Alte Maße, Münze und Gewichte* (Mannheim: Bibliographisches Institut/Mayers Lexikonverlag, 1987), 77. I have seen no evidence in the documentary record in Stuttgart for the story Kopp relates about Honauer hiding a boy in a trunk, who secretly threw gold into the crucible during the process. Hermann Kopp, *Die Alchemie in älterer und neuerer Zeit*, 2 vols. (Heidelberg: Carl Winter's Universitätsbuchhandlung, 1886), 165–81. Vladimir Karpenko repeats this tale in his "The Chemistry and Metallurgy of Transmutation," *Ambix* 39, no. 2 (1992): 49.

28. Kopp is the most prominent example of this view: Kopp, *Die Alchemie in älterer und neuerer Zeit*, 158–84. See also Katrin Cura, "Die Alchemisten und das Gold: Echte und falsche Alchemisten, ihre Laboratorien und Laboranten," *Kultur & Technik* 3 (1998): 34–41; and Bruce T. Moran, *Distilling Knowledge: Alchemy, Chemistry, and the Scientific Revolution* (Cambridge, Mass.: Harvard University Press, 2005), 30.

29. William R. Newman, *Promethean Ambitions: Alchemy and the Quest to Perfect Nature* (Chicago: University of Chicago Press, 2004).

30. Robert Halleux, "L'alchemiste et l'essayeur," in *Die Alchemie in der europäischen Kultur- und Wissenschaftsgeschichte*, ed. Christoph Meinel (Wiesbaden: Harrassowitz, 1986), 277–92; William R. Newman, "The Place of Alchemy in the Current Literature on Experiment," in *Experimental Essays: Versuche zum Experiment*, ed. Michael Heidelberger and Friedrich Steinle (Baden-Baden: Nomos, 1998), 9–33; William R. Newman and Lawrence M. Principe, *Alchemy Tried in the Fire: Starkey, Boyle, and the Fate of Helmontian Chymistry* (Chicago: University of Chicago Press, 2002), chap. 2.

31. Newman and Principe, *Alchemy Tried in the Fire*, 41.

32. Vannoccio Biringuccio, *La pirotechnia* (Venice: Venturino Roffinello. Ad instantia di Curio Nau. & Fratelli, 1540); Georgius Agricola, *De re metallica libri XII* (Basel: Froben, 1556); Lazarus Ercker, *Beschreibung allerfürnemsten mineralischen Ertzt unnd Berckwercksarten* (Prague: Georg Schwartz, 1574). For a clear explanation of several of these methods, which were used in the laboratory in Oberstockstall/Kirchberg am Wagram, see Soukup and Mayer, *Alchemistisches Gold,* esp. chap. 5.

33. Lazarus Ercker, *Treatise on Ores and Assaying,* trans. and ed. Annelise Grünhaldt Sisco and Cyril Stanley Smith (Chicago: University of Chicago Press, 1951), xv.

34. See Soukup and Mayer, *Alchemistisches Gold,* 95–119.

35. *The pirotechnia of Vannoccio Biringuccio: The Classic Sixteenth-Century Treatise on Ores and Assaying,* ed. and trans. Cyril Stanley Smith and Martha Teach Gnudi, rev. ed. (1959; facsimile, New York: Dover Publications, 1990), 159–60.

36. Soukup and Mayer, *Alchemistisches Gold,* 95–96.

37. Duke Friderich to Georg Gader, 19 August 1596, HStA Stuttgart, Bü. 1, Nr. 7.

38. *The pirotechnia of Vannoccio Biringuccio,* ed. and trans. Smith and Gnudi, 201. On antimony, see Claus Priesner and Karin Figala, eds., *Alchemie: Lexikon einer hermetischen Wissenschaft* (Munich: Beck, 1998), s.v. "Antimon."

39. According to Cyril Stanley Smith, "[Lazar] Ercker's treatment [of assaying] is not basically different from that found in any early twentieth-century book on fire assaying. Analysis in the sixteenth-century sense differs from modern chemical analysis mainly in that a pure metal or an elemental substance (rightly or wrongly so called) had to be separated and weighed. The fixed mass ratio between the elements and the compounds which the modern analyst weights because they are easily separable was not recognized for two centuries." Introduction to Ercker, *Treatise on Ores and Assaying,* xi–xii.

40. Bruce Moran offers a very good, succinct discussion of why alchemical transmutation was both "possible and justified" in early modern Europe in his article "The Alchemist's Reality: Problems and Perceptions of a German Alchemist in the 17th Century," *Halcyon: A Journal of the Humanities* 9 (1987) 133–48. For another take on the problem of understanding early modern rationality on its own terms, see Darren Oldridge, *Strange Histories: The Trial of the Pig, the Walking Dead, and Other Matters of Fact from the Medieval and Renaissance Worlds* (London: Routledge, 2005).

41. Draft of a contract between Duke Friedrich and Dr. Johann Hofrichter, ca. 1600, HStA Stuttgart, Bestand A47, Bü. 4, Nr. 6. As he also explained in this letter, however, he did not believe that manufactured tinctures could accomplish this; rather, it required "as the only natural beginning the true primordial [*ursprünglich*] seed of such a tincture, without which one can come to neither a good middle nor end."

42. Moran also points out the "powerful intellectual-mystical justification for the claims of alchemical enthusiasts" found in the notion of the transmutation

of bread and wine into the body and blood of Christ in the Catholic Mass. Moran, "The Alchemist's Reality," 135.

43. Ibid., 135.

44. Karpenko, "The Chemistry and Metallurgy of Transmutation," 49.

45. Ibid., 50.

46. Ibid.

47. Ibid., 51.

48. Karpenko's terminology. Ibid., 49.

49. Philip Benedict discusses these kinds of contracts in his "Faith, Fortune and Social Structure in Seventeenth-Century Montpellier," *Past and Present* 152 (1996): 46–78.

50. Carol L. Loats, "Gender and Work in Paris: The Evidence of Employment Contracts, 1540–1560," *Proceedings of the Western Society for French History* 20 (1993): 25–37.

51. On painters' contracts in fifteenth-century Italy see Michael Baxandall, *Painting and Experience in Fifteenth-Century Italy* (Oxford: Oxford University Press, 1972), 3–27. On the construction trade in Renaissance Florence, see Richard A. Goldthwaite, *The Building of Renaissance Florence: An Economic and Social History* (Baltimore: Johns Hopkins University Press, 1980). On contracts between patients and healers in early modern Bologna, see Gianna Pomata, *Contracting a Cure: Patients, Healers, and the Law in Early Modern Bologna* (Baltimore: Johns Hopkins University Press, 1998), esp. chap. 2. See also Hannelore Glasser, *Artists' Contracts of the Early Renaissance*, Outstanding Dissertations in the Fine Arts (New York: Garland Publishing, 1977), 5–10. Glasser discusses the emergence of the "work contract" from Roman law, and its distinction from a wage contract.

52. On Vřesovic, see Václav Březan, *Životy Posledních Rožmberků*, ed. Jaroslav Pánek, 2 vols. (Prague: Svoboda Praha, 1985), 283, 84, 701; Evans, *Rudolf II and His World*, 216.

53. Vřesovec to Rožmberk, 13 December 1576, SOA Třeboň RRA 25. As cited and translated in Vladimír Karpenko, "Bohemian Nobility and Alchemy in the Second Half of the 16th Century: Wilhelm of Rosenberg and Two Alchemists," *Cauda Pavonis: The Hermetic Text Society Newsletter* 15, no. 2 (1996): 16.

54. Rožmberk appears in this document by the German version of his name, Wilhelm Rosenberg.

55. This translation is based loosely on Vladimír Karpenko's in Karpenko, "Bohemian Nobility and Alchemy in the Second Half of the 16th Century," 14–18. The original agreement between Claudius Syrrus Romanus and Vilém Rožmberk, dated 9 January 1577, can be found in SOA Třeboň RRA 25.

56. This correspondence is located in HStA Stuttgart, Bestand A47, Bü. 4, Nr. 5.

57. "I cannot have less than three months," he added; "I will not tarry long." Johann Hofrichter to Duke Friedrich of Württemberg, 29 December 1600, HStA Stuttgart, Bestand A47, Bü. 4, Nr. 6.

58. Pomata, *Contracting a Cure*, 28.

59. Baxandall, *Painting and Experience in Fifteenth-Century Italy,* 5.

60. Pomata, *Contracting a Cure,* 28.

61. On Keller's contract with Marx Fugger in 1570, see Soukup and Mayer, *Alchemistisches Gold,* 30 n. 69. Keller also worked for Elector August of Saxony in the "Goldhaus" and could well have been the same Daniel Keller who later worked for Duke Friedrich of Württemberg; see chap. 5, below, and Rudolf Kötzschke and Hellmut Kretzschmar, *Sächsische Geschichte* (Frankfurt am Main: Verlag Wolfgang Weidlich, 1965), 222, where Kretzschmar claims that "Dr. Daniel Keller of Augsburg was particularly entrusted with the work [of the Goldhaus]."

62. On Polhaimer, see Jost Weyer, "Der 'Goldmacher' Michael Polhaimer—Alchemistischer Betrüger am Hof des Grafen Wolfengang II. von Hohenlohe," *Beitrage zur Landeskunde: Regelmäßige Beilage zum Staatsanzeiger für Baden-Württemberg* 4 (1993): 7–11, and *Graf Wolfgang II. von Hohenlohe und die Alchemie,* 229 and, for a transcription of the original, 482–83.

63. Hereward Tilton, *The Quest for the Phoenix: Spiritual Alchemy and Rosicrucianism in the Work of Count Michael Maier (1569–1622),* ed. Cristoph Markschies and Gerhard Mueller, Arbeiten zur Kirchengeschichte 88 (Berlin: Walter de Gruyter, 2003), 70–86.

64. This distinction finds parallels in other trades as well. Artists, for example, could work either by selling individual paintings or (more rarely) by obtaining appointments as court painters; likewise, some medical practitioners worked as court physicians, while empirics hawked cures in the marketplace.

65. Duke Friedrich, for instance, promised Johann Hofrichter "princely protection [*Schutz und Schirm*] from all unreasonable force." Agreement between Johann Hofrichter and Duke Friedrich, ca. 1600, HStA, Bestand A47, Bü. 4, Nr. 6.

66. For an overview of Müller/von Mühlenfels's life and career, see Priesner and Figala, *Alchemie,* s.v. "Müller v. Mühlenfels"; and Weyer, *Graf Wolfgang II. von Hohenlohe und die Alchemie,* 310. On Grumbach, see chap. 3. Other alchemists who received castles, estates, or noble titles include Conrad Schuler, who received the Epstein estate in Hessen, and Hans Heinrich Nüschler, who received the *Freihof* of Kirchheim (equipped with a laboratory).

67. This contract is quoted in Karl von Weber, *Aus vier Jahrhunderten: Mittheilungen aus dem Haupt-Staatsarchive zu Dresden* (Leipzig: Verlag von Bernhard Tauchnitz, 1858), 2:17.

68. "Verpflichtung des Michael Heinrich Wagenmann vom Hoff, Herzog Friedrich gegen 4000 fl [i.e., gulden] die Hohe Theophrastische Medizin des Universals zuwege zu bringen," 23 December 1598, HStA Stuttgart, Bü. 3, Nr. 6.

69. Theodor Wagner certainly saw it this way, as did Henry Carrington Bolton. Wagner, "Wissenschaftlicher Schwindel aus dem südlichen Böhmen," *Mittheilungen des Vereins für Geschichte der Deutschen in Böhmen* 16 (1878): 112–23; Bolton, *The Follies of Science at the Court of Rudolph II, 1576–1612* (Milwaukee: Pharmaceutical Review Publishing Co., 1904).

70. Polhaimer's process could transform two pounds (or sixty-four loth) of

quicksilver into ten loth of fine silver. Weyer has calculated that in this period one pound of quicksilver cost on average one gulden, and one pound of silver cost twenty gulden. Therefore, if Polhaimer and Wolfgang bought one pound of quicksilver for one gulden, they would have gotten out of it three gulden, eight kreuzer worth of silver (one gulden = sixty kreuzer). Weyer, *Graf Wolfgang II. von Hohenlohe und die Alchemie*, 230–31.

71. See her manuscript on the philosophers' stone, 1 April 1573, NStAW, 1 Alt 9, Nr. 306; as well as Tara E. Nummedal, "Alchemical Reproduction and the Career of Anna Maria Zieglerin," *Ambix* 48 (2001): 56–68.

72. On vitriol/*Kupferwasser* (copper or iron sulfate), see Weyer, *Graf Wolfgang II. von Hohenlohe und die Alchemie*, 175.

73. "Verzeichnis der dem Honauer liefernden materialien," 31 August 1596, HStA Stuttgart, Bü. 1, Nr. 7. One württembergische zentner = 46.77 kg. See Hans-Georg Hofacker, "*. . . sonderlich hohe Künste und vortreffliche Geheimnis": Alchemie am Hof Herzog Friedrichs I. von Württemberg—1593 bis 1608.* (Stuttgart: Verein der Freunde des Chemischen Instituts Dr. Flad e. V., 1993), 64. The meaning of "Venetian soap [*Seife*]" in this context is unclear. The phrase could simply refer to an olive-oil soap; however, Grimm's Wörterbuch offers a subsidiary definition of *Seife* as a colloquial miner's term for a mineral deposit that appears on the surface of the earth after it is disturbed to open up a mine. On this definision of *Seife*, see *Deutsches Wörterbuch von Jacob Grimm und Wilhelm Grimm* (Leipzig: S. Hirzel, 1854–1960).

74. Agreement between Claudius Syrrus Romanus and Vilém Rožmberk, 9 January 1577, SOA Třeboň RRA 25.

75. Pomata, *Contracting a Cure*, 28.

76. Ibid., 42.

CHAPTER FIVE

1. Sister Virginia Heines, ed., *Libellus de Alchimia, Ascribed to Albertus Magnus* (Berkeley: University of California Press, 1958), 12. (Note, however, that this text is now thought to be pseudonymous.)

2. Howard Louthan and Andrea Sterk, eds., *John Comenius: The Labyrinth of the World and the Paradise of the Heart* (New York: Paulist Press, 1998), 114.

3. Owen Hannaway, "Laboratory Design and the Aim of Science," *Isis* 77 (1986): 585–610; William R. Newman, "Alchemical Symbolism and Concealment: The Chemical House of Libavius," in *The Architecture of Science*, ed. Peter Galison and Emily Thompson (Cambridge, Mass.: MIT Press, 1999), 59–78; Jole Shackelford, "Tycho Brahe, Laboratory Design, and the Aim of Science: Reading Plans in Context," *Isis* 84 (1993): 211–30.

4. In recent years historians increasingly have paid attention to the relationship between the sites of scholarship (such as laboratories, museums, and studies) and the types of knowledge produced in these spaces. See, for example, B. J. T.

Dobbs, "From the Secrecy of Alchemy to the Openness of Chemistry," in *Solomon's House Revisited: The Organization and Institutionalization of Science*, ed. Tore Frängsmayr (Stockholm, Sweden: Science History Publications, 1989), 75–94; Paula Findlen, "Masculine Prerogatives: Gender, Space, and Knowledge in the Early Modern Museum," in Galison and Thompson, *The Architecture of Science*, 29–57; Deborah E. Harkness, "Managing an Experimental Household: The Dees of Mortlake and the Practice of Natural Philosophy," *Isis* 88, no. 2 (1997): 247–62; and Steven Shapin, "The House of Experiment in Seventeenth-Century England," *Isis* 79 (1988): 373–404.

5. William R. Newman and Lawrence M. Principe, *Alchemy Tried in the Fire: Starkey, Boyle, and the Fate of Helmontian Chymistry* (Chicago: University of Chicago Press, 2002), 96. For some general comments on the problematic use of laboratory images as transparent representations of real laboratories, see C. R. Hill, "The Iconography of the Laboratory," *Ambix* 22, no. 2 (1975): 102–10. On the images accompanying Petrarch's and Brant's texts, as well as Brueghel's engraving, see chap. 2. For an overview of the seventeenth-century Netherlandish genre paintings of alchemists, see Lawrence Principe and Lloyd DeWitt, *Transmutations: Alchemy in Art* (Philadelphia: Chemical Heritage Foundation, 2002). Heinrich Khunrath's well-known image of a "laboratorium-oratorium" in his *Amphitheatrum sapientia aeternae* (Hanau: G. Antonius, 1609) has also been the subject of several studies, including Ralf Töllner, *Der unendliche Kommentar: Untersuchungen zu vier ausgewählten Kupferstichen aus Heinrich Khunraths "Amphitheatrum sapientiae aeternae solius verae" (Hanau 1609)* (Ammersbek bei Hamburg: Verlag an der Lottbek, 1991); and, most recently, Peter Forshaw, "'Alchemy in the Amphitheatre': Some Considerations of the Alchemical Content of the Engravings in Heinrich Khunrath's *Amphitheatre of Eternal Wisdom* (1609)," in *Art and Alchemy*, ed. Jacob Wamberg (Copenhagen: Museum Tusculanum Press, 2006), 195–220.

6. R. W. Soukup, S. von Osten, and H. Mayer, "Alembics, Cucurbits, Phials, Crucibles: A 16th-Century Docimastic Laboratory Excavated in Austria," *Ambix* 40, no. 1 (1993), 25; Rudolf Werner Soukup and Helmut Mayer, *Alchemistisches Gold, Paracelsistische Pharmaka*, ed. Helmuth Grössing, Karl Kadletz, and Marianne Klemun, Perspektiven der Wissenschaftsgeschichte, vol. 10 (Vienna: Böhlau Verlag, 1997); Marcos Martinón-Torres and Thilo Rehren, "Alchemy, Chemistry and Metallurgy in Renaissance Europe: A Wider Context for Fire-Assay Remains," *Historical Metallurgy* 39, no. 1 (2005): 14–28; Jost Weyer, *Graf Wolfgang II. von Hohenlohe und die Alchemie: Alchemistische Studien in Schloß Weikersheim, 1587–1610*, ed. Historischen Verein für Württembergisch Franken, Stadtarchiv Schwäbisch Hall and Hohenlohe-Zentralarchiv Neuenstein, Forschungen aus Württembergisch Franken, vol. 39 (Sigmaringen: Jan Thorbecke Verlag, 1992).

7. Newman and Principe, *Alchemy Tried in the Fire*, 3, 96, and entire.

8. Duke Friedrich instructed Osiander, "We graciously wish you not to delay in installing your noble and our trusted Georg Honauer, Lord zu Brunnhoff and

Grabenschutz, in a number of rooms in the Alte Lusthaus in our zoological garden (because he plans to run trials and set up [*probieren und errichten*] a number of things)." Duke Friedrich to Lucas Osiander the Younger, 3 September 1596, HStA Stuttgart, Bestand A47, Bü 1, Nr. 7. See also Duke Friedrich to Baumeister Heinrich Schickhardt, 3 September 1596, HStA Stuttgart, Bestand A47, Bü 1, Nr. 7. For further information on Heinrich Schickhardt, see Werner Fleischhauer, *Renaissance im Herzogtum Württemberg* (Stuttgart: W. Kohlhammer, 1971), 276ff.

9. From 1669 the building was used as the ducal *Kunstkammer,* housing the Kunst- and Altertumskabinett founded by Duke Friedrich. For a discussion of this building's architecture and its place in the ducal gardens, see Stefan Gugenhan, *Die landesherrlichen Gärten zu Stuttgart im 16. und 17. Jahrhundert* (Stuttgart: Klett-Cotta, 1997), 76–84. See also Hans-Georg Hofacker, "*. . . sonderlich hohe Künste und vortreffliche Geheimnis": Alchemie am Hof Herzog Friedrichs I. von Württemberg—1593 bis 1608.* (Stuttgart: Verein der Freunde des Chemischen Instituts Dr. Flad e. V., 1993), 44–56.

10. For a modern facsimile of the festival book that contained this engraving, see Ludwig Krapf and Christian Wagenknecht, eds., *Stuttgarter Hoffeste: Texte und Materialien zur höfischen Repräsentation im frühen 17. Jahrhundert: Esaias van Hulsen/Matthäus Merian, Representatio der Furstlichen Aufzug und Ritterpsil: Die Kupferstichfolge von 1616* (Tübingen: Max Niemeyer Verlag, 1979).

11. For a description of the architecture of these *Lusthäuser* in the context of other central European *Lusthäuser,* see Gugenhan, *Die landesherrlichen Gärten,* 76ff.

12. Many of the architectural details of what follows are drawn from ibid.

13. Ibid., 84.

14. Describing Libavius's critique of Tycho's laboratory at Uraniborg, Jole Shackelford notes: "A key component in Libavius's characterization of Tycho Brahe's contemplative Paracelsian science is his portrayal of Uraniborg as a place of darkness: darkness surrounds Tycho as he looks to the heavens from his upper-story observatory, darkness envelops the chemical research conducted in the basement laboratory, and this darkness reaches out to taint the morality of Uraniborg and its inhabitants. The darkness that envelops Tycho's science in Pythagorean secrecy is connected with the dark recesses of the earth, home to the forces of evil." Shackelford, "Tycho Brahe, Laboratory Design, and the Aim of Science," 213.

15. On *Essen* and their use in alchemy, see Weyer, *Graf Wolfgang II. von Hohenlohe und die Alchemie,* 128.

16. "Inventarium aller transmutir- und Medicinal stück, wie auch Rohen, undt praeparierten Materialien, Instrumenten, Oefen, Gläser, Tigel, Treibscherben, undt anderer zum laborieren requirierten Werckzeug, so sich nicht allein in alten Lusthaus, sondern auch andern laboratorijs im Schloßgartten gefunden . . . ," 28 January–3 February 1608 (unpaginated), HStA Stuttgart, Bestand A47, Bü. 9, Nr. 9.

17. See Hofacker, "*. . . sonderlich hohe Künste,*" 46–47.

18. On *Scheidekolbe,* see Soukup and Mayer, *Alchemistisches Gold,* 115–16;

Soukup von Osten, and Mayer, "Alembics, Cucurbits, Phials, Crucibles"; Weyer, *Graf Wolfgang II. von Hohenlohe und die Alchemie,* 135.

19. Weyer, *Graf Wolfgang II. von Hohenlohe und die Alchemie,* 137.

20. For a good overview of the distillation techniques and equipment associated with alchemy, see Soukup and Mayer, *Alchemistisches Gold,* 121–72.

21. As Soukup and Mayer have argued recently, "Without question, docimasy, namely, the analysis of ores and alloys of gold, silver, [and] copper, had great significance in Oberstockstall. . . . For anyone who had to examine ores or metal for coins, as well as for those who wanted to know whether their transmutations were finally crowned with success, exact balances, precise observations, and accurate computation were more important than wishes, conjectures, theories, and images." Ibid., 2.

22. For a description of the process of refining silver with *Testscherben,* see ibid., 85–88.

23. Mühlenfels was employed as an alchemist in Stuttgart until he was executed in 1606. Claus Priesner and Karin Figala, *Alchemie: Lexikon einer hermetischen Wissenschaft* (Munich: Beck, 1998), s.v. "Müller v. Mühlenfels, Johann." For his career in Stuttgart, see HStA Stuttgart, Bestand A47, Bü. 8 and 9.

24. See "Inventarium aller transmutir- und Medicinal stück."

25. On touchstones and needles, see Lazar Ercker, *Beschreibung Allerfürnemsten Mineralischen Ertzt unnd Berckwercksarten* (Prague: Georg Schwartz, 1574); and Weyer, *Graf Wolfgang II. von Hohenlohe und die Alchemie,* 142.

26. On the utility of antimony and arsenic in alchemical processes, see Priesner and Figala, *Alchemie,* s.v. "Antimon" and "Arsen."

27. Hofacker summarizes this information succinctly and explains in detail the uses of each of these materials in Hofacker, *". . . sonderlich hohe Künste,"* 45–47.

28. For an imaginary stereotypical laboratory scene, see Newman and Principe, *Alchemy Tried in the Fire,* 35.

29. "Inventarium aller transmutir- und Medicinal stück."

30. Georgius Butina, for instance, was listed as "ein vertrybener Pfarherr aus der Steuermarck," Kyliannus Hofmann as a doctor, and Pantaleon Keller as "ein Balbirer undt assistant." Ibid. (unpaginated).

31. Hans-Georg Hofacker calculated that Friedrich employed a total of forty-three *Laboranten* to work in the Alte Lusthaus during his reign. Hofacker, *". . . sonderlich hohe Künste,"* 52.

32. "Statt und Ordnung der Goldmacher," n.d. [1596], HStA Stutgart, Bü. 1, Nr. 7. For a discussion of this document, see ibid., 50–51.

33. "Statt und Ordnung der Goldmacher."

34. It is unclear whether there were two keys to the trunk, or two to the building. "The two of them should have each other's keys [*abgewechselte Schlussel*], so that neither can get in without the other." Ibid.

35. Daniel Keller "hat allein, was im bevelch von Ihr f. G. ihn ufferlegt worden laborirt." Ibid. [unpaginated]. Interestingly, a Daniel Keller apparently also signed

an alchemical contract with Marx Fugger in 1570. Soukup and Mayer, *Alchemistisches Gold,* 30 n. 69.

36. Hofacker points this out in *". . . sonderlich hohe Künste,"* 33.

37. Duke Friedrich of Württemberg borrowed the services of this "particular well-known artisan, called the great Jewish artisan from Ferrara" from Emperor Rudolf II in 1596. Writing to Rudolf, Friedrich explained that he had heard about this man, a Ferrarese Jew named Abraham Calorne who could make saltpeter out of earth. Although saltpeter was used in some alchemical processes, its production was not always alchemical per se (except, perhaps, in the very loosest sense of transmutating earth into saltpeter); nonetheless, Calorne's process offered the possibility that Württemberg could become a major supplier of saltpeter, which was a crucial ingredient in gunpowder and thus in high demand in early modern Europe. As "a prince who loves the arts," Friedrich asked Emperor Rudolf's permission to "possibly address himself to the above-mentioned artisan." Moreover, Duke Friedrich wondered whether Rudolf might grant Calorne six weeks leave to come to Stuttgart. Rudolf evidently complied, for Calorne arrived in Stuttgart shortly thereafter to begin his work. Duke Friedrich I to Kaiser Rudolf II, 22 December 1596, HStA Stuttgart, Bü. 3, Nr. 11. On the production of saltpeter and its importance in early modern central Europe, see Priesner and Figala, *Alchemie,* s.v. "Salpeter"; and Weyer, *Graf Wolfgang II. von Hohenlohe und die Alchemie,* 328–38.

38. "Inventarium aller transmutir- und Medicinal stück."

39. Ibid.

40. Ibid. The former pastor Georgius Butina, who worked in the Alte Lusthaus from 1602 until 1608, "is also a bound assistant, [and] at the moment is working on his own process, with the approval of Your Grace." Ibid.

41. Christoph Fabri (1596–1601) and Paul Meier (1600–1606). See Walter Pfeilsticker, *Neues Württembergisches Dienerbuch* (Stuttgart: J. G. Cotta'sche Buchhandlung, 1957–63), vol. 1, §1841.

42. The "Bosselknechte" (tinkers' assistants) employed during Friedrich's reign included Sebastian Brecht (1596–99), Heinrich Eckhardt/Eckert (1596–1608), Melchior Hu(o)ber (1576–77), Thomas Nallinger (1605–6), and Peter Schellkopf (1605–8). "Bossler" (tinkers) included Johannes Geissler (1607–8) and Matthäus Schmid (1606–7). The only two "Hafner" (potters) listed were Christoff Wagner (1597–1608) and Hippolyt (Bolay) Reinhardt (1597/98–1620). See ibid., §§1839–44; and also Hofacker, *". . . sonderlich hohe Künste,"* 52.

43. "Inventarium aller transmutir- und Medicinal stück."

44. Pfeilsticker, *Neues Württembergisches Dienerbuch,* vol. 1, §1843.

45. Although we know almost nothing about Scheinpflugerin, her fellow alchemists' comments suggest that she had a certain authority in the Třeboň laboratory and may indeed have taken a leading role in managing the alchemical experiments conducted there. Two alchemists writing to Rožmberk, Hans Härpfell and Johannes de Sole, both refer to having heard word through Salome about their alchemical assignments. In de Sole's case, the unfortunate news that Wilhelm had

withdrawn his support came from "my sister Salome" (a sorore mea Salome). SOA Třeboň RRA 25.

46. "Rewers Florian Capplers, Inspektor der *Laboranten* im Fürstliche Lustgarten, dem Herzog treu und gehorsam zu sein," 20 June 1597, HStA Stuttgart, Bü. 3, Nr. 2.

47. These quantities were measured in Scheffel and Eimer: 6 Scheffel of rye, 24 Scheffel of spelt, 4 Scheffel of oats, and 4 Eimer of wine. Ibid. The other *Inspektor,* Christoff Wagner, received a similar salary: 110 gulden yearly, 28 Scheffel Dinkel, 5 Schefel rye, 4 Scheffel oats, four Eimer wein (1 Eimer = 2, 940 hl), 2 yearly outfits, ten pounds of candles, and 6 Stamess of wood "from the woodpile." "Inventarium aller transmutir- und Medicinal stück."

48. Athanasius Kircher, *Mundus subterraneus,* vol. 2 (Amsterdam: J. Janssen, 1678), 321, as cited and translated in Martha Baldwin, "Alchemy and the Society of Jesus in the Seventeenth Century: Strange Bedfellows?" *Isis* 40 (1993): 49–50.

49. "Inventarium aller transmutir- und Medicinal stück."

50. Steven Shapin makes this point about experimental spaces in general through the late seventeenth century, in "The House of Experiment in Seventeenth-Century England," 377.

51. See Daniel Prandtner von Prandt to Vilém Rožmberk, 4 December 1576, and Claudius Syrrus Romanus's contract of 9 January 1577, SOA Třeboň RRA 25. George Starkey offers a seventeenth-century example. See Newman and Principe, *Alchemy Tried in the Fire,* 96–100. Robert Boyle also had laboratories in his three homes, as did his employee Godfrey Hanckwitz. See Shapin, "The House of Experiment in Seventeenth-Century England," 379–80; and Lawrence M. Principe, *The Aspiring Adept: Robert Boyle and His Alchemical Quest* (Princeton, N.J.: Princeton University Press, 1998), 134–37.

52. Shapin, "The House of Experiment in Seventeenth-Century England," 393.

53. Deborah Harkness demonstrates this most clearly in her analysis of the Dees' house at Mortlake in her "Managing an Experimental Household," 247–62.

54. Shackelford, "Tycho Brahe, Laboratory Design, and the Aim of Science," 217.

55. Weyer, *Graf Wolfgang II. von Hohenlohe und die Alchemie,* 100–101.

56. On the experimental spaces associated with the Royal Society, see Shapin, "The House of Experiment in Seventeenth-Century England"; Steven Shapin and Simon Schaffer, *Leviathan and the Air-Pump: Hobbes, Boyle, and the Experimental Life* (Princeton, N.J.: Princeton University Press, 1985), 36. For a more recent overview of the emergence of laboratories as experimental spaces, see Pamela H. Smith, "Laboratories," in *The Cambridge History of Science,* vol. 3, *Early Modern Europe,* ed. Lorraine Daston and Katharine Park (Cambridge: Cambridge University Press, 2006), 289–305. I am grateful to Pamela Smith for sharing this essay with me in draft form.

57. Although a great deal of alchemy was situated at princely courts, it is important to recognize that laboratories rarely served as theaters of spectacle and

courtly *sprezzatura* in the same way that the collections of Italian naturalists did. Unlike natural magic, anatomies, or experiments, transmutations and other alchemical processes were long, laborious, dirty processes that were ill suited for courtly games. Paula Findlen has drawn attention to the experimental culture associated with collections of natural objects in early modern Italy. Paula Findlen, *Possessing Nature: Museums, Collecting, and Scientific Culture in Early Modern Italy* (Berkeley: University of California Press, 1994), chap. 5. Marco Bragadino's transmutations in Venice before the doge and other political officials are an important exception. Ivo Striedinger, *Der Goldmacher Marco Bragadino* (Munich: Theodor Ackerman, 1928).

58. For another example of the importance of the "industrial" context of mining in alchemy, see William R. Newman, *Gehennical Fire: The Lives of George Starkey, an American Alchemist in the Scientific Revolution* (Cambridge, Mass.: Harvard University Press, 1994); and Newman and Principe, *Alchemy Tried in the Fire*, 157–61.

59. Pamela H. Smith, *The Body of the Artisan: Art and Experience in the Scientific Revolution* (Chicago: University of Chicago Press, 2004).

60. On painters' workshops, see Melissa Meriam Bullard, "Heroes and Their Workshops: Medici Patronage and the Problem of Shared Agency," *Journal of Medieval and Renaissance Studies* 24 (1994): 179–98.

61. William R. Newman, "Alchemy, Assaying, and Experiment," in *Instruments and Experimentation in the History of Chemistry*, ed. Frederic L. Holmes and Trevor Levere (Cambridge, Mass.: MIT Press, 2000), 35, and "The Place of Alchemy in the Current Literature on Experiment," in *Experimental Essays: Versuche zum Experiment*, ed. Michael Heidelberger and Friedrich Steinle (Baden-Baden: Nomos, 1998).

62. Lawrence M. Principe, "Apparatus and Reproducibility in Alchemy," in Holmes and Levere, *Instruments and Experimentation in the History of Chemistry*, 71.

63. Rudolf Kötzschke and Hellmut Kretzschmar, *Sächsische Geschichte* (Frankfurt am Main: Verlag Wolfgang Weidlich, 1965), 221.

64. "Inventarium über das Goldhauß," 1598, SHStA Dresden, Geheimer Rat (Geheimes Archiv), Loc 4419/3.

65. Ibid., fols. 23–30.

66. Ibid., fols. 30ff.

67. Ibid.

68. Ibid., fol. 35v.

69. Ibid., fol. 38.

70. Ibid., fols. 33–34.

71. Ibid., fol. 34r–v

72. Libavius, however, placed his distillation apparatus inside his sanctuary, unlike the arrangement in the Dresden Goldhaus.

73. Jost Weyer has painstakingly reconstructed this laboratory in his *Graf*

Wolfgang II. von Hohenlohe und die Alchemie. On Count Wolfgang's correspondence with Friedrich on alchemical matters, see 304–9.

74. For a description of the interior of the laboratory, see ibid., 96–103. For an English summary, see Smith, "Laboratories."

75. Weyer, *Graf Wolfgang II. von Hohenlohe und die Alchemie,* 320–26.

76. On saltpeter production in Weikersheim, see ibid., 328–38.

77. On secrecy in alchemy, see Newman, *Gehennical Fire,* 62–78; Newman and Principe, *Alchemy Tried in the Fire,* 179–86; and Principe, *The Aspiring Adept,* chap. 5, and "Robert Boyle's Alchemical Secrecy: Codes, Ciphers and Concealments," *Ambix* 39 (1992): 63–74. On secrecy in natural knowledge more generally, see William Eamon, *Science and the Secrets of Nature: Books of Secrets in Medieval and Early Modern Culture* (Princeton, N.J.: Princeton University Press, 1994); and Pamela O. Long, *Openness, Secrecy, Authorship: Technical Arts and the Culture of Knowledge from Antiquity to the Renaissance* (Baltimore: Johns Hopkins University Press, 2001).

78. Newman, "Alchemical Symbolism and Concealment," 59–60.

79. William Eamon has articulated the multiple valences of secrecy in the sixteenth century as "esoteric wisdom, the domain of occult or forbidden knowledge, the artisan's cunning, the moral injunctions to protect secrets from the *vulgus,* and the political power that attended knowledge of secrets." Eamon, *Science and the Secrets of Nature,* 5.

80. Jakob Faber to Vilém Rožmberk, Tuesday after Lichtmesse [5 February] 1585, SOA Třeboň RRA 25.

81. The author of *Libellus de Alchimia,* for instance, maintained that "no one should begin operations without plenty of funds, so that he can obtain everything necessary and useful for this art." Heines, *Libellus de Alchimia,* 14. See also Michael Maier, *Examen fucorum pseudo-chymicorum* (Frankfurt am Main: Theodor de Bry, 1617), reprinted in Wolfgang Beck, "Michael Maiers Examen Fucorum Pseudo-Chymicorum: Eine Schrift wider den falschen Alchemisten" (PhD diss., Fakultät für Chemie, Biologie und Geowissenschaft der Technischen Universität München, 1992), 12.

82. Contract between Claudius Syrrus Romanus and Vilém Rožmberk, 9 January 1577, SOA Třeboň RRA 25.

83. Meister Peter Hottenstein to Herzog Friedrich, 28/29 December 1597, HStA Stuttgart, Bestand A47, Bü. 3, Nr. 4.

84. Reuers Florian Capplers, 20 June 1597, HStA Stuttgart, Bestand A47, Bü. 3, Nr. 2.

85. Gadi Algazi, "Scholars in Households: Refiguring the Learned Habitus, 1480–1550," *Science in Context* 16, no. 1/2 (2003): 9–42; Findlen, "Masculine Prerogatives"; Harkness, "Managing an Experimental Household"; Shapin, "The House of Experiment in Seventeenth-Century England."

86. Newman, "Alchemical Symbolism and Concealment," 71.

87. Ibid. On dispersion, see also Newman and Principe, *Alchemy Tried in the Fire,* 186–87.

88. Newman, "Alchemical Symbolism and Concealment," 71.

89. Tycho Brahe, *Tychonis Brage Dani Opera Omnia,* ed. J. L. E. Dreyer, H. Raeder, and E. Nystrom, vol. 5 (Copenhagen: Gyldendal, 1913–29), 117–18, as quoted in Hannaway, "Laboratory Design and the Aim of Science," 598.

90. Shackelford, "Tycho Brahe, Laboratory Design, and the Aim of Science," 227.

CHAPTER SIX

1. Pfalzgraf Richard zu Simmern to Herzog Julius, 29 December 1586, NStA Wolfenbüttel, 1 Alt 9, Nr. 213, fol. 11v.

2. See Jost Weyer, *Graf Wolfgang II. von Hohenlohe und die Alchemie: Alchemistische Studien in Schloß Weikersheim, 1587–1610,* ed. Historischen Verein für Württembergisch Franken, Stadtarchiv Schwäbisch Hall and Hohenlohe-Zentralarchiv Neuenstein, Forschungen aus Württembergisch Franken, vol. 39 (Sigmaringen: Jan Thorbecke Verlag, 1992), 228–71.

3. For a description of this horribly botched execution, see the report from the Fugger newsletters reprinted in Ivo Striedinger, *Der Goldmacher Marco Bragadino* (Munich: Theodor Ackerman, 1928), 337–38.

4. For very brief summaries of all of these cases see Weyer, *Graf Wolfgang II. von Hohenlohe und die Alchemie,* 272–319.

5. For an overview of the early modern penal system in the Holy Roman Empire, see Richard van Dülmen, *Theater des Schreckens: Gerichtspraxis und Strafrituale in der frühen Neuzeit,* 3rd ed. (Munich: Beck, 1988).

6. William R. Newman, *Promethean Ambitions: Alchemy and the Quest to Perfect Nature* (Chicago: University of Chicago Press, 2004), 54–62. Jean-Jaques Magnet collected and printed excerpts from a number of these texts in the early eighteenth century. Ioh. Chrysippo Faniano, "De Iure Artis Alchemiae, hoc est, Variorum Authorum & praesertim Iruisconsultorum Iudicia & Responsa ad Quaestionem, An Alchemia sit Ars legitima? colligente Ioh. Chrysippo Faniano," in *Bibliotheca chemica curiosa,* ed. Jean-Jaques Manget (Geneva: sumpt. Chouet, G. De Tournes, Cramer, Perachon, Ritter, & S. De Tournes, 1702; reprint, [Bologna?]: Arnoldo Forni Editore, 1976), 210–16 in 1976 ed. See also J. R. Partington, "Albertus Magnus on Alchemy," *Ambix* 1, no. 1 (1937): 13–15; Lynn Thorndike, *A History of Magic and Experimental Science* (New York: Columbia University Press, 1934–58), 3:48–50.

7. Lynn Thorndike, for example, briefly described the arguments of the fourteenth-century figures John Andrea, Andrea de Rampoinis of Isernia, and Alberico da Rosciate of Bergamo, who seem to be more concerned with the selling of alchemical gold than the theological status of alchemical claims to transmutation. Lynn Thorndike, *A History of Magic and Experimental Science,* 3:48–50;

as well as Newman, *Promethean Ambitions,* 54–62 and 91–97. Despite William Newman's important explication of the connections between alchemy and demonology, the full import of writings in canon law with respect to alchemy are still poorly understood and would benefit greatly from further study.

8. The full Latin text of this decretal, as well as an English translation, can be found in Partington, "Albertus Magnus on Alchemy," 15–16. I have relied on Partington's translation here.

9. On European "Geldpolitik" around 1300 and the issue of debased coinage, see Barbara Obrist, "Die Alchemie in der mittelalterlichen Gesellschaft," in *Die Alchemie in der europäishen Kultur- und Wissenschaftsgeschichte,* ed. Christoph Meinel, Wolfenbütteler Forschungen, Band 32 (Wiesbaden: Otto Harrassowitz, 1986), 49–51.

10. Partington, "Albertus Magnus on Alchemy," 16.

11. These authors' opinions on the legal status of alchemy would also merit further study. For a brief overview of their views, see Lynn Thorndike, *A History of Magic and Experimental Science,* 3:50; Will H. L. Ogrinc, "Western Society and Alchemy from 1200–1500," *Journal of Medieval History* 6 (1980): 109; Ioh. Chrysippo Faniano, "De Iure Artis Alchemiae," 201–16.

12. On these two laws, see Ogrinc, "Western Society and Alchemy," 119.

13. Obrist, "Die Alchemie in der mittelalterlichen Gesellschaft," 49–52.

14. Dorothea Waley Singer, *Catalogue of Latin and Vernacular Alchemical Manuscripts in Great Britain and Ireland, Dating from before the XVI Century,* 3 vols. (Brussels: M. Lamertin, 1928), 3:777–78, as cited in Obrist, "Die Alchemie in der mittelalterlichen Gesellschaft," 51 n. 89.

15. D. Geoghegan, "License of Henry VI to Practise Alchemy," *Ambix* 6 (1957): 10–17; Ogrinc, "Western Society and Alchemy," 119–20; F. Sherwood Taylor, *The Alchemists, Founders of Modern Chemistry* (New York: H. Schuman, 1949), 123–44.

16. On princes' condemnation of false alchemical coinage and their simultaneous reliance on alchemists to solve fiscal crises, see Bruce T. Moran, *Distilling Knowledge: Alchemy, Chemistry, and the Scientific Revolution* (Cambridge, Mass.: Harvard University Press, 2005), 31–34; Obrist, "Die Alchemie in der mittelalterlichen Gesellschaft," 49–52; Ogrinc, "Western Society and Alchemy," 117–23; Jost Weyer, "Der Alchemist im lateinischen Mittelalter (13. bis 15. Jahrhundert)," in *Der Chemiker im Wandel der Zeiten: Skizzen zur geschichtlichen Entwicklung des Berufbildes,* ed. Eberhard Schmauderer and Gesellschaft Deutscher Chemiker, Fachgruppe Geschichte der Chemie (Weinheim: Verlag Chemie, 1973), 38–40.

17. Reichsstadt Nürnberg, Mandate 1696 (Verbot der Alchemie), 12 December [1493], Staatsarchiv Nürnberg, Reichsstadt Nürnberg, Fürstentum Ansbach, Archivakten Nr. 318. Reprinted in Willy Bein, *Der Stein der Weisen und die Kunst Gold zu machen: Irrtum und Erkenntnis in der Wandlung der Elemente, mitgeteilt nach den Quellen der Vergangenheit und Gegenwart,* vol. 88, *Voigtländers Quellenbücher* (Leipzig: Voigtländer, ca. 1915), 97.

18. The 1493 Nuremberg law seems to have had more bark than bite, as the city archive yields no evidence of an individual actually bringing charges under this law. Correspondence of the author with Dr. Peter Fleischmann, Staatsarchiv Nürnberg, 3 November 1998.

19. Counterfeiting coins could involve anything from stamping a false insignia on them to using bad metal or improper weights. Gustav Radbruch, ed., *Peinliche Gerichtsordnung Kaiser Karls V. von 1532 (Carolina),* ed. Arthur Kaufmann, Universal-Bibliothek, no. 2990 (Stuttgart: Philipp Reclam, 1996), articles 111–14.

20. The sick in Bologna, for instance, could appeal to the Protomedicato when they felt that healers had not fulfilled their obligations to heal, while Florentines could appeal to the mason's guild to mediate disputes over construction contracts. Richard A. Goldthwaite, *The Building of Renaissance Florence: An Economic and Social History* (Baltimore: Johns Hopkins University Press, 1980), 142–43; Gianna Pomata, *Contracting a Cure: Patients, Healers, and the Law in Early Modern Bologna* (Baltimore: Johns Hopkins University Press, 1998), chap. 2.

21. The archival documents do not indicate the outcome of this case. Ober- and Untervogt zu Kirchheim to Herzog Friedrich, 28 June 1603, HStA Stuttgart, Bestand A47, Bü. 6, Nr. 1. Kopp, however, cites Karl Pfaff's claim that Stocker was imprisoned. Karl Pfaff, *Die Geschichte Wirtembergs,* I Abth., 240ff., as cited in Hermann Kopp, *Die Alchemie in älterer und neuerer Zeit,* 2 vols. (Heidelberg: Carl Winter's Universitätsbuchhandlung, 1886), 1:181.

22. Friedrich to Georg Gadner, 19 August 1596; Friedrich to Landschreiber Erhard Stickel, 18 September 1596; Friedrich to Honauer, 9 October 1596; and Friedrich to Erhard Stickel, 9 October 1596; all collected in HStA Stuttgart, Bestand A47, Bü. 1, Nr. 7.

23. The figure of two hundred thousand ducats appears in a written report from the imperial goldsmith, Hans Müller, about his contact with Honauer over the years. He claimed that Honauer wrote about the duke's desire to see two hundred thousand ducats in a letter. See Hans Müller, "Verzeichnis, was ich ungefehr acht jahren vom Georg Honauer von Olmutz der sich Philosophia Alchamia schreiben thuet waiß und erfahren, auch sich zum theil zuegetragen, 13 February 1597," HStA Stuttgart, Bestand A47, Bü. 1, Nr. 10, (pt. 2). Muller's letter also describes a transmutation that Honauer successfully carried out at a military camp in Hungary, the gold from which, according to the assays, was good. Honauer cites the other two figures—seven thousand gulden and the thirty-six thousand ducats monthly—in an appeal for help to Emperor Rudolf II written after he was imprisoned. Honauer to Emperor Rudolf II, 5 January 1597, HStA Stuttgart, Bestand A47, Bü. 1, Nr. 10 (pt. 1).

24. According to a *Decret* from 12 October 1596, which appears to be no longer extant. The archivist Güntzler's summary of the case from 1823, however, notes this document. Güntzler's summary of the Honauer case, November 1823, HStA Stuttgart, Bestand A47, Bü. 2, Nr. 9.

25. "Extract aus dem peinlichen Gerichtsprotokoll," 3 April, 1597, HStA

Stuttgart, Bestand A47, Bü. 2, Nr. 8. Figure 12, as well as a second woodcut commemorating Honauer's execution, are reprinted in Walter L. Strauss, *The German Single-Leaf Woodcut, 1550–1600: A Pictorial Catalogue* (New York: Abaris Books, 1975), 2:497 and 3:1352.

26. Friedrich to Moritz of Hessen-Kassel, 19 January 1597, HStA Stuttgart, Bestand A47, Bü. 1, Nr. 10 (pt. II).

27. "I worked day and night," he wrote. Wagenmann to Duke Friedrich, ca. 1599, HStA Stuttgart, A47, Bü. 3, Nr. 6.

28. This suggestion is in Friedrich Nick, *Stuttgarter Chronik und Sagenbuch: Eine Sammlung denkwürdiger Begebenheiten, Geschichten und Sagen der Stadt Stuttgart und ihrer Gemarkung.* (Stuttgart: Emil Gutzkow, Verlagsbuchhandlung, 1875), 217–19. See also Walter Pfeilstecker, *Neues Württembergisches Dienerbuch* (Stuttgart: Cotta'sche Buchhandlung, 1957–63), §1839 and §590.

29. See chap. 4.

30. The details of Reiche's case, including the additional charges against him regarding illegitimate children with his concubine Appolonia and a Bürger's daughter named Beatrix Nussin, are located in HStA Stuttgart, A47, Bü. 5 (entire).

31. Hans Heinrich Nüschler to Duke Friedrich I of Württemberg, 15 July 1601 (Nüschler's plea for mercy), HStA Stuttgart, Bestand A47, Bü. 3, Nr. 10.

32. Ibid.

33. At this time, I have not uncovered cases of alchemists' bringing other alchemists to trial for *Betrug;* however, it would be fascinating to see if further archival research could uncover such cases.

34. Alten Vogt zu Kirchheim, Christoff Haan, to Rat Johann Kielmann, 20 February, 1608, HStA Stuttgart, Bestand A47, Bü. 3, Nr. 10.

35. Ewald Hoghelande, *Kurtzer Bericht und klarer Beweiss,* trans. Joachim Tancke (Leipzig: M. Lantzenberger for J. Apel, 1604), 5; Latin version as Ewaldus de Hoghelande, *Historiae aliquot transmutationis metallicae scriptae pro defensione alchymiae contra hostium rabiem* (Coloniae Agrippinae: 1604).

36. Andreas Libavius, *Alchemia Andreae Libavii* (Frankfurt am Main: Excudebat J. Saurius impensis Petri Kopffii, 1597), "Lectori salutem," VIII; for a modern German translation, see Andreas Libavius et al., *Die Alchemie des Andreas Libavius: Ein Lehrbuch der Chemie aus dem Jahre 1597, zum ersten mal in deutscher Übersetzung mit einem Bild- und Kommentarteil* (Weinheim: Verlag Chemie, 1964), x.

37. Libavius, *Alchemia,* "Lectori salutem," VIII; Libavius et al., *Die Alchemie des Andreas Libavius,* x.

38. [Johann Valentin Andreae], *Chymische Hochzeit Christiani Rosencreutz. Anno 1459. Arcana publicata vilescunt, & gratiam prophanata amittunt. Ergo: ne Margaritas obÿceporcis, seu Alsino substerne rosas* (Strasbourg: Lazarus Zetzner, 1616), reprinted in *Fama fraternitatis (1614), Confessio fraternitatis (1615), Chymische Hochzeit Christiani Rosencreutz: Anno 1459 (1616),* ed. Richard van Dülmen (Stuttgart: Calwer Verlag, 1973), 279–95. I have used the English translations here

from Adam McLean, ed., *The Chemical Wedding of Christian Rosenkreutz,* trans. Joscelyn Godwin (Grand Rapids, Mich.: Phanes Press, 1991).

39. Andreae's clergyman father Johann Andreae (1554–1601) was a practicing alchemist in the duchy of Württemberg, and his mother Maria (1550–1632) became court apothecary to Duke Friedrich in Tübingen when her husband died in 1601. Andreae's brother, a pastor, apparently spent much of his life in alchemical pursuits as well. Donald R. Dickson, "Johann Valentin Andreae's Utopian Brotherhoods," *Renaissance Quarterly* 49 (1996): 763–64.

40. McLean, *The Chemical Wedding,* 33.

41. Ibid., 40.

42. Ibid., 43–44.

43. Ibid., 44.

44. Ibid., 47.

45. Ibid., 48.

46. For another alchemical text involving a judicial metaphor, see Johann Sternhals, *Ritter Krieg; das ist, Ein philosophisch Gedicht, in Form eines gerichtlichen Process* ([Erffordt]: Martin Wittel, 1595), in which Gold and Iron both argue their virtues before the judge, Mercury. Sömmering recommended this text to Duke Julius of Braunschweig-Wolfenbüttel (see chap. 1).

47. Michael Maier, *Examen fucorum pseudo-chymicorum* (Frankfurt am Main: Theodor de Bry, 1617), reprinted in Wolfgang Beck, "Michael Maiers Examen Fucorum Pseudo-Chymicorum: Eine Schrift wider den falschen Alchemisten" (PhD diss., Fakultät für Chemie, Biologie und Geowissenschaft der Technischen Universität München, 1992), 10 in 1617 ed.

48. Maier, *Examen fucorum pseudo-chymicorum,* 7–8.

49. On this frontispiece, see Beck, "Michael Maiers Examen Fucorum Pseudo-Chymicorum," 23–26.

50. *Res publicae,* of course, can mean either the state or the public good, so it is not entirely clear whether Maier situates himself in the same line of thought as legal scholars who saw false alchemy and counterfeiting as a danger to the state, or whether he sees it as a danger to society more generally. In his German translation of the *Examen,* Wolfgang Beck interprets *res publicae* to mean "state": "Da solche Menschen sowohl für den Staat als auch für die Wissenschaft der Chemie sehr schädlich sind." Maier, *Examen fucorum pseudo-chymicorum,* 10; German translation in Beck, "Michael Maiers Examen Fucorum Pseudo-Chymicorum," 72. Interestingly, Maier felt that the name "alchemy" had been so besmirched by this second group, the drones, that he preferred not to use it for the "true" art that he espoused; he used the term *Alchymista,* therefore, only in the most negative sense, as a synonym for drone, or *Betrüger,* reserving the term *Chymicus* for the true practitioners. One should not make too much of this distinction between the *pseudo-chymici* and the drone/*Alchymisti,* however, for Maier did not always apply it consistently. He conflated the two throughout his treatise, most obviously in the title of the treatise, which purports to be an examination of the "pseudo-

chemical drones" (*Examen fucorum pseudo-chymicorum*). Moreover, the subtle difference between a *pseudo-chymicus* and a drone was far less important to Maier than that between the entire cluster of *pseudo-chymici* drones and alchemists and the true *chymici.* See, for instance, "Eodem modo in hac arte magnus est proventus Pseduo chymicorum fucorum." On the issue of terminology, see Beck, "Michael Maiers Examen Fucorum Pseudo-Chymicorum," 74 n. 270.

51. For more on Maier's views on these points, see chap. 1.

52. Maier, *Examen fucorum pseudo-chymicorum,* p. 22. Maier even cites the civil code in support of this point: "The civil laws allow those engaged in trade to bargain with each other when they are buying and selling, but not to charge or pay more than one and a half times the true worth. . . . But should one be defrauded out of more than half the appropriate price, then both buyer and seller are entitled to redress. . . . *See the section* [of the civil law] *on legal claims regarding buying and selling.* In fact, with the merchandise of the pseudochymical drones [*fucorum Pseudo-chymicorum*], nothing is related, because they provide words and false promises for gold, but how much empty wind one must sell for a drachma of gold cannot be established." Maier, *Examen fucorum pseudo-chymicorum,* 22.

53. Ibid., 23.

54. Literally "owls," a reference to Greek coins, on which the owl, the symbol of Athens, was stamped. See Beck, "Michael Maiers Examen Fucorum Pseudo-Chymicorum," 74 n. 269.

55. Maier, *Examen fucorum pseudo-chymicorum,* 11. The Paracelsian Leonhard Thurneisser invoked a similar image in 1583 when he wrote that false alchemists "wear large chains, and broad armor studded with gold, and daggers of white copper or some other mixture, also exquisite clothing and golden rings." Leonhard Thurneisser, *Magna alchymia* (Berlin: Niclaus Voltz, 1583), unfoliated.

56. Andreas Libavius too decried what he called "the presumptuous arrogance and boasting of certain people." Libavius et al., *Die Alchemie des Andreas Libavius,* x.

57. [Andreae], *Fama fraternitatis (1614), Confessio fraternitatis (1615), Chymische Hochzeit Christiani Rosencreutz: Anno 1459 (1616),* 60. Translation based on McLean, *The Chemical Wedding,* 31. Andreae would have had reason to perceive some alchemists as dangerous frauds. As Donald Dickson has argued, Andreae felt that his lifelong dream "to found a brotherhood of Christians dedicated to ameliorating society's ills," articulated in the *Chymical Wedding* and Rosicrucian manifestoes, was severely damaged when the idea of such a society "became associated in the public eye with vulgar alchemy and sectarian lunacy." See Dickson, "Johann Valentin Andreae's Utopian Brotherhoods," 760–802.

58. Maier, *Examen fucorum pseudo-chymicorum,* 28.

59. Nor did they need flashy equipment, according to Alexander Lauterwald: "Du darffst kein grossn verlag darzu / Und kansts machen mit guter ruh / Viel öffn und glase laß nicht machn / Eins dient wol zu diesen sachen / Darinnen wird das edle Gut / Gemacht davon man sagen thut" (You must not devote much to it

/ And [you] can do it very calmly / It is not necessary to have many furnaces and glasses made / One will do fine for this / In which will be made the precious treasure / Of which much has been said). Alexander Lauterwald, *Colloquium Philosophicum* (Cologne: Heinrich Netessem, 1597), D3.

60. Maier, *Examen fucorum pseudo-chymicorum,* 29.

61. Ibid.

62. Ibid., 30.

63. On Maier's lifelong efforts to secure patronage for his work, see Hereward Tilton, *The Quest for the Phoenix: Spiritual Alchemy and Rosicrucianism in the Work of Count Michael Maier (1569–1622),* ed. Cristoph Markschies and Gerhard Mueller, Arbeiten zur Kirchengeschichte 88 (Berlin: Walter de Gruyter, 2003); and Karin Figala and Ulrich Neumann, "'Author cui nomen Hermes Malavici': New Light on the Bio-Bibliography of Michael Maier (1569–1622)," in *Alchemy and Chemistry in the 16th and 17th Centuries,* ed. Piyo Rattansi and Antonio Clericuzio (Dordrecht: Kluwer Academic, 1994), 121–47.

64. *Via veritatis, das ist: Ein wahrhafftiger philosophischer Bericht, aus dem rechten und wahren Fundament der Natur genommen, und den irrended Alchimisten . . . von einem Liebhaber der Warheit an tag gegeben* (n.p., 1635), "Ad Lectorem," I, ij.

65. Thrasybulus Ricenus [Heinrich Khunrath], "Treuhertzige Warnungs-Vermahnung eines Getreuen Liebhabers der Warheit an alle wahre Liebhaber der Naturgemässen Alchymiae Transmutatoriae; daß wegen der Bübischen Handgriffe der betriegerischen Arg-Chymisten gute Auffacht vonnöthen," in *Von hylealischen, das ist pri-materialischen Catholischen oder allgemeinen natürlichen Chaos: Der naturgemässen Alchymiae und Alchymisten* (Magdeburg, 1597; reprint, Graz: Akademische Druck- u. Verlagsanstalt, 1990), 268–86.

66. On transmutation histories, see William R. Newman, *Gehennical Fire: The Lives of George Starkey, an American Alchemist in the Scientific Revolution,* 2nd ed. (Chicago: University of Chicago Press, 2003), 3–13; and Lawrence M. Principe, *The Aspiring Adept: Robert Boyle and His Alchemical Quest* (Princeton, N.J.: Princeton University Press, 1998), 93–98, 108–11.

67. Maier, *Examen fucorum pseudo-chymicorum,* 31.

68. Ibid., 35.

69. Ibid., 33–34.

70. This distinction, of course, harkens back to earlier debates about human technology and whether it was possible to imitate nature. Newman, *Promethean Ambitions.*

71. Anna Zieglerin's book on the preparation of the philosophers' stone, 1 April 1573, NStA Wolfenbüttel, 1 Alt 9, Nr. 306, fol. 68.

72. In her study of practitioners of witchcraft and magic in London in the 1980s, anthropologist Tanya Luhrmann describes a similar situation regarding beliefs about the prevalence of black magic. "I never encountered anything remotely resembling Satanism, though there were rumors. . . . Indeed, magicians seem very

concerned about morality. They talk about black magic; they usually tell you that there are black magicians elsewhere and stress that they, by contrast, are very white. Inevitably there are unstable individuals who lay claim to evil powers, but in my fieldwork I met no-one of that ilk who was not personally isolated and obviously mentally disturbed, and there were enough of those. Throughout my work I met no group, nor any stable individual, who actually seemed to engage in practices which other magicians—or indeed, the wider public—would call 'black.' Black magic seemed to be a myth, and the talk about it seemed to be part of a general determination to be as morally virtuous as possible." T. M. Luhrmann, *Persuasions of the Witch's Craft: Ritual Magic in Contemporary England* (Cambridge, Mass.: Harvard University Press, 1989), 81.

73. Philipp Sömmerings Verhör und Aussage, 9 July 1574, NStA Wolfenbüttel, 1 Alt 9, Nr. 309.

74. Ducal Hofrat G. Wirtemberger to Duke Wilhelm V, 23 April 1591. Striedinger, *Der Goldmacher Marco Bragadino,* no. 348.

75. Georg Honauer to Emperor Rudolf II, 5 January 1597, HStA Stuttgart, Bestand 47, Bü. 1, Nr. 10 (pt. II).

76. Draft of a contract between Duke Friedrich and Dr. Johann Hofrichter, ca. 1600, HStA Stuttgart, Bestand A47, Bü. 4, Nr. 6.

SELECTED BIBLIOGRAPHY

ARCHIVAL AND MANUSCRIPT COLLECTIONS

Niedersächsisches Staatsarchiv Wolfenbüttel (NStA Wolfenbüttel)
Bestand Ältere Landesakten (Alt)
1 Alt (Das alte Fürstliche Hauptarchiv)
9 (Acta Publica aus der Regierungszeit des Herzog Julius 1568–89)
Nr. 158, 160 (Korrespondenz des Landgrafen Wilhelms IV von Hessen mit Julius)
Nr. 213 (Korrespondenz des Pfalzgrafen Richard zu Simmern 1580–82)
Nr. 306–36 (Die Untersuchung gegen die betrügerischen Goldmacher)
2 Alt (Kanzelei, Geheime Ratstube)
24 (Die Neigung des Herzogs Julius zur Alchimie 1571–85)
Nr. 67 (Ratesprotokoll über Unterredung mit den Alchimsten Mortz Lam und Georg von Minden [1576], Der von Kasper Uden empfohlene jüdische Alchimist Aaron Goldschmidt [1576])
Nr. 2211 (Verhör des auf Veranlassung Philipp Sömmerings heimlich nach Wolfenbüttel gekommenen Hermann Sprenger)
Sächsisches Haupstaatsarchiv Dresden (SHStA Dresden)
Geheimer Rat/Geiheimes Archiv, Loc. 4416/6–4419/19 (Alchemistische Sachen)
Württembergisches Hauptstaatsarchiv Stuttgart (HStA Stuttgart)
Bestand A47, Büschel 1–9 (Akten über Alchemisten im Dienst von Herzog Friedrich I. von Württemberg 1595–1615)
Státní oblastní archiv Třeboň (SOA Třeboň)
Rožmberkský roddiný archiv 25 (Familie Rosenberg) 25, Briefe 1528–96 (Natürliche und geheime Wissenschaften)
Rukopisy Třeboň, skupina C

Bayerisches Hauptstaatsarchiv München (BHStA Munich)
Fürstensachen 426A, Fol. 46, 55–57, 65, 93–96b, 97–99, 102, 186, 282, 327

PRINTED PRIMARY SOURCES

Agricola, Georgius. *De re metallica* (German vernacular edition). Basel: Froben, 1561.

———. *De re metallica libri XII*. Basel: Froben, 1556.

Agrippa von Nettesheim, Heinrich Cornelius. *De incercitudine et vanitate scientiarum atque artium declamatio*. Antwerp: Grapheus Drucker, 1530.

[Andreae, Johann Valentin]. *Chymische Hochzeit Christiani Rosencreutz. Anno 1459. Arcana publicata vilescunt, & gratiam prophanata amittunt. Ergo: ne Margaritas obÿceporcis, seu Alsino substerne rosas*. Strasbourg: Lazarus Zetzner, 1616.

August, Elector of Saxony. *Künstlich Obst Garten Büchlein: Churfürst Augusti zu Sachsen . . . Jtzo auffs New von einem Liebhaber deß Gartenbawes wiederumb an Tag gegeben*. Magdeburg: Betzel, 1620; Berlin: Martin Guth, 1636.

Bergwerk- und Probierbüchlein: A Translation from the German of the "Bergbüchlein," a Sixteenth-Century Book on Mining Geology and of the "Probierbüchlein," a Sixteenth-Century Work on Assaying. Edited by Annelise Grünhaldt Sisco and Cyril Stanley Smith. New York: American Institute of Mining and Metallurgical Engineers, 1949.

Bernard of Trevisan, Denis Zacaire, Nicholas Flamel, and Gerhard Dorn. *De chymico miraculo, quod lapidem philosophiae appellant*. Basel: Ex officina haeredum Petri Pernae, 1583.

Bernhardus Trevisanus. *Morieni Romani. Item, primum in Lucem prodit Bernardi Trevirensis Responsio ad Thomam*. Paris: Gulielimus Guillard, 1564.

Beuther, David. *Davids Beuthers . . . Universal, und vollkommener Bericht, von der hochberümbten Kunst der Alchymj und seinen in solcher erlangten, und erkundigten Secreten, und Kunststücklein . . . Sampt beygefügtem Gespräch, von Betrug und Irrweg, etlicher unerfahrnen Laboranten, so sich betrieglich vor Alchymisten dargeben . . . Ex bibliotheca chymica D. Iohannis Ernesti Burggrauii*. Frankfurt: In Verlegung Wilhelm Fitzers, 1631.

Biringuccio, Vannoccio. *La pirotechnia*. Venice: Venturino Roffinello, Ad instantia di Curio Nau. & Fratelli, 1540.

Bodin, Jean. *Respublica das ist: Gründtliche und rechte Underweysung, oder eigentlicher Bericht, in welchem aussführlich vermeldet wirdt, wie nicht allein das Regiment wol zubestellen, sonder auch in allerly Zustandt, so wol in Krieg unnd widerwertigkeit, als Frieden und Wolstand zuerhalten zey . . . Jetzt . . . auss lateinischer unnd frantzösischer Sprach, in unser . . . Teutsch*. Translated by Johann Oswaldt. Mumpelgart: Jacques Foillet, 1592.

Brant, Sebastian. *Das Narrenschyf*. Basel: Johann Bergmann von Olpe, 1494.

———. *The Ship of Fools, by Sebastian Brant.* Edited and translated by Edwin H. Zeydel. New York: Columbia University Press, 1944.

———. *Stultifera navis.* Translated by Jacob Locher. Basel: Johann Bergmann de Olpe, 1497.

Březaň, Václav. *Životy Posledních Rožmberků.* Edited by Jaroslav Pánek. 2 vols. Prague: Svoboda Praha, 1985.

Caesar, Theophilus, ed. *Alchimyspiegel: oder Kurtz entworffene Practick, der gantzen Chimischen Kunst: neben Anzeig, welche darzu tüglich seyen, oder nit: Wie der Alten mit seltzamen verdunckelten Reden und Wörtern hievon beschrieben Bücher zuverstehen: Und darinnen sonderlich der falschen Alchimisten Betrug entdecket wird. Alles in zweyen lustigen Gesprächen verfasset: Und das erste vor diesem aufs dem Arabischen von Roberto Castrensi in Latein, Nun aber sampt dem andern so newlich Lateinische beschriben worden, in unser Teutsche Sprach ubergesetzt, Durch Teophilum Caesarem August.* Frankfurt am Main: Vincentii Steinmeyer, 1597.

Cellini, Benvenuto. *The Treatises of Benvenuto Cellini on Goldsmithing and Sculpture.* Translated by C. R. Ashbee. Bow, England: Laurence Hodson, 1898; reprint, New York: Dover Books, 1967.

Chaucer, Geoffrey. *The Canterbury Tales: A Verse Translation.* Translated by David Wright. Oxford: Oxford University Press, 1985.

Clajus, Johann. *Altkumistica, das ist: Die Kunst, aus Mist durch seine Wirckung, Gold zu machen: Wider die betrieglichen Alchimisten, und ungeschickte vermeinte Theophrasisten.* Leipzig: Z. Bärwaldt, 1586.

Coelum philosophorum; seu, Secreta naturae. Paris: Vivantius Gaultherot, 1543.

Dorn, Gerhard, and Adam von Bodenstein. *Dictionarium Theophrasti Paracelsi/ Onomasticon Theophrasti Paracelsi.* Frankfurt: Christoff Rab, 1584.

Ebener, Erasmus. "Bericht an Herzog Julius von Braunschweig vom 26. Januar 1572. Mit mineralogischen, metallurgischen und chemischen Anmerkungen vom Zehntner Meyer und vom Bergamts-Auditor Hausmann." *Hercynisches Archiv* (Halle) 1, no. 1–4 (1805): 494–540.

Epimentheus, Franciscus [Hieronymus Reusner]. *Pandora, Das ist, Die Edleste Gab Gottes, oder der Werde unnd Heilsamme Stein der Weisen, mit welchem die alsten Philosophi auch Theophrastus Paracelsus, die unvolkomene Metallen, durch gewalt des Fewrs verbessert: Sampt allerley schädliche und unheisame Kranckheiten, innerlich und eusserlich haben vertrieben.* Basel: [S. Apario], 1582.

Erasmus, Desiderius. *De utilitate colloquiorum.* Basel: Johannes Froben, 1526.

Ercker, Lazarus. *Beschreibung allerfürnemsten mineralischen Ertzt unnd Berckwercksarten.* Prague: Georg Schwartz, 1574.

Faniano, Ioh. Chrysippo. "De Iure Artis Alchemiae, hoc est, Variorum Authorum & praesertim Iruisconsultorum Iudicia & Responsa ad Quaestionem, An Alchemia sit Ars legitima? colligente Ioh. Chrysippo Faniano." In *Biblio-*

theca chemica curiosa, edited by Jean-Jaques Manget, 210–16. Geneva: sumpt. Chouet, G. De Tournes, Cramer, Perachon, Ritter, & S. De Tournes, 1702.

Gratarolo, Guglielmo, ed. *Alchemiae, quam vocant, artisque metallicae*. Basel: Henricus Petri & Petrus Perna, 1572.

———, ed. *Verae alchemiae artisque metallicae citra aenigmata, doctrina, certusque modus, scriptis tum nouis tum ueteribus nunc primùm & fideliter maiori ex parte editis, comprehensus: quorum elenchum à praefatione reperies*. Basel: Henricus Petri & Petrus Perna, 1561.

Herliche medicische [sic] Tractat, vor nie in Truck kommen. Edited by Heinrich Wolff, Giovanni Braccesco, Dominicus Burgauer, Wolff Geuss, and Johann Vogt. Strasbourg: Bernhart Jobin, 1576.

Hoghelande, Ewald. *Historiae aliquot transmutationis metallicae scriptae pro defensione alchymiae contra hostium rabiem*. Coloniae Agrippinae, 1604.

———. *Kurtzer Bericht und klarer Beweiss: das die Alchimey, oder . . . die Goldmacherkunst, ein sonderbar Geschenck Gottes*. Translated by Joachim Tancke. Leipzig: M. Lantzenberger for J. Apel, 1604.

Hoghelande, Theobald. *De alchemiae difficultatibus . . . : In quo docetur quid scire, quidque vitare debeat verae chemiae studiosus . . . et multae philosophorum propositiones obscurae et difficiles explicantur*. Cologne: H. Falckenburg, 1594.

———. *Von den Irrwegen der Alchimisten: Das ist: Wolbedenckliche Erinnerung welcher gestalt vor der unerfahrnen unnd vermeinten Alchimsiten ublichen unnd doch nichts nützigen Prackticken neben dero hinderlichen Zufällen, so ihnen auss manglung der Kunst und Verstands täglichen obligen, sich männiglichen zu verhüten habe . . . Erstlich durch Theobaldum de Hohenlande Middelburg in Lateinischer Sprach beschrieben, nun mehr aber . . . in Hochteutsche Sprach . . . versetzet Durch Johannem Hippodamum Cheruscum*. Frankfurt am Main: Johann Spiess, 1600.

Huser, Johannes. *Erster[-zehender] Theil der Bücher und Schriften . . . jetzt auffs neu . . . an Tag geben . . . .* 10 vols. Basel: Conrad Waldkirch, 1589–90.

[Kertzenmacher, Peter.] *Alchimia das ist alle Farben, Wasser, Olea, Salia, und Alumina, damit man alle Corpora Spiritus unnd Calces prepariert, sublimiert unnd fixiert zubereyten: und wie man diese Ding nutze, auff dass Sol und Luna werden möge: auch von Soluieren unnd Scheydung aller Metall, Polierung allerhandt Edelgestein, fürtrefflichen Wassern zum etzen, scheyden unnd soluieren: und zuletzt wie die gifftige Dämpff zuuerhüten, ein kurtzer Bericht*. Frankfurt am Main: Bey C. Engenoffs Erben, 1570.

———. *Alchimi und Bergwerck. Wie alle Farben, Wasser, Olea, Salia, und Alumina damit man alle Corpora Spiritus und Calces preparirt, sublimirt und fixirt gemacht sollen werden lert das erst Büch: das ander Büch zeygt an wie man dise ding nütze auff das Sol und Luna werden mög und vom Soluiren auch Scheydung aller Metal Polirung allerhandt Edelgesteyn fürtreflichen

Wassern zum etzen, scheyden und soluieren, und zületst wie die gifftie Dämpff der Metal zu verhüten auss Archilae, Koleno, &c. eyn kurtzer Begreiff. Strasbourg: Jacon Cammerlander, 1534.

Khunrath, Heinrich. *Von hylealischen, das ist pri-materialischen Catholischen oder allgemeinen natürlichen Chaos: Der naturgemässen Alchymiae und Alchymisten.* Magdeburg, 1597; reprint, Graz: Akademische Druck- u. Verlagsanstalt, 1990.

Kieser, Franz, Alexander von Suchten, and Georgius Clettus. *Cabala chymica; Concordantia chymica; Azot philosoph. solificatum drey unterschiedliche, nützliche, und zuvor nie aussgegangene Tractätlein, ohn welcher Hülff niemandt in Ewigkeit Chymiam veram verstehen, noch das summum arcanum erlernen wirdt.* Mülhausen: bey Martin Spiessen in Verlegung Johann Spiessen &c., 1606.

Lauterwald, Alexander. *Colloquium Philosophicum. Von der warenn Chimia, Sapientia, und Natura rerum, Wie die von menniglichen und allen Liebhabern der Kunst von aller Sophisterey und betriegery mag unterschieden und verstanden werden.* Cologne: Heinrich Netessem, 1597.

———. *Widerlegung der Altkuhmisterey, Die M. Johannes Claius Pfarherr zu Bengellegeb, aus seiner OxKudomia ausgelassen, wider die hochlöbliche verborgene Geheimnis, der Natürlichen warhafftigen Kunst der Alchymiae, die gedachter Claius in genere verwirfft, und felschlich für lauter Betriegerey helt, Allen Liebhabern der waren Alchemiae zu geschrieben.* Cologne: Nettesheim, 1597.

Libavius, Andreas. *Alchemia Andreae Libavii.* Frankfurt am Main: Excudebat J. Saurius impensis Petri Kopffii, 1597.

Lull, Ramón. *De alchimia opuscula quae sequuntur.* Nuremberg: Johannes Petreius, 1546.

———. *Libelli aliquot chemici.* Basel: Peter Perna, 1572.

———. *Ramundii Lullii opera.* Strasbourg: Lazar Zetzner, 1598.

Maier, Michael. *Examen fucorum pseudo-chymicorum.* Frankfurt am Main: Theodor de Bry, 1617.

———. *Atalanta Fugiens.* Oppenheim: Theodor de Bry, 1618.

Norton, Thomas. *The Ordinall of Alchimy.* In *Theatrum chemicum britannicum containing severall poeticall pieces of our famous English philosophers, who have written the hermetique mysteries in their owne ancient language,* edited by Elias Ashmole. London: Printed by J. Grismond for Nath. Brooke, 1652; reprint, Baltimore: Williams and Wilkins Co., 1031.

Paracelsus. *Sämtliche Werke, 1. Abteilung: Medizinische, naturwissenschaftliche und philosophische Schriften.* Edited by Karl Sudhoff. 14 vols. Munich: R. Oldenbourg, 1922/23.

Petrarca, Francesco. *Von der Artzney bayder Glück, des guten und widerwertigen.* Augsburg: Heinrich Stayner, 1532.

Radbruch, Gustav, ed. *Peinliche Gerichtsordnung Kaiser Karls V. von 1532 (Caro-*

lina). Edited by Arthur Kaufmann. Universal-Bibliothek, no. 2990. Stuttgart: Philipp Reclam, 1996.

Rhenanus, Johannes. *Das deutsche Salinenwesen im 16. Jahrhundert: Reiseberichte des Allendorfer Salzgrafen Johannes Rhenanus (um 1528–1589).* Edited by Hans-Henning Walter. Freiberg: Bergakademie Freiberg, 1989.

Ricenus, Thrasybulus [Heinrich Khunrath]. "Treuhertzige Warnungs-Vermahnung eines Getreuen Liebhabers der Warheit an alle wahre Liebhaber der Naturgemässen Alchymiae Transmutatoriae; daß wegen der Bübischen Handgriffe der betriegerischen Arg-Chymisten gute Auffacht vonnöthen." In *Vom hylealischen, das ist pri-materialischen Catholischen oder allgemeinen natürlichen Chaos: Der naturgemässen Alchymiae und Alchymisten.* Magdeburg, 1597; reprint, Graz: Akademische Druck- u. Verlagsanstalt, 1990.

Severinus, Petrus. *Idea medicinae philosophicae, fundamenta continens totius doctrinae Paracelsicae, Hippocraticae, & Galenicae.* Basel: Sixtus Henricpetrus, 1571.

Sternhals, Johann. *Ritter Krieg; das ist, Ein philosophisch Gedicht, in Form eines gerichtlichen Process, wie zwey Metallen, nemlich, Sol und Mars durch Klag, Antwort, und Beweisz, jegliches Natur und Eygenschafft von jrem natürlichen Gott und Richter Mercurio gehöret, und entlich durch ein wolgegründtes Urtel, mit ewigwerender Freundtschafft einig zusamen verbunden werden.* [Erffordt]: Martin Wittel, 1595.

Thurneisser zum Thurm, Leonhard. *Magna alchymia.* Berlin: Niclaus Voltz, 1583.

———. *Melitsah kai hermâeneia = das ist, Onomasticum und Interpretatio, oder aussfuhrliche Erklerung.* Berlin: Nicolas Voltzen, 1574–83.

Toxites, Michael. *Onomastica II.* [Strasbourg: Per Bernhardum Jobinum], 1574.

Via veritatis, das ist: Ein wahrhafftiger philosophischer Bericht, aus dem rechten und wahren Fundament der Natur genommen, und den irrended Alchimisten . . . von einem Liebhaber der Warheit an tag gegeben. N.p., 1635.

SECONDARY SOURCES

Algazi, Gadi. "Food for Thought: Hieronymus Wolf Grapples with the Scholarly Habitus." In *Egodocuments and History: Autobiographical Writing in Its Social Context since the Middle Ages,* edited by Rudolf Dekker, 21–44. Hilversum: Verloren, 2002.

———. "Gelehrte Zerstreutheit und gelernet Vergeßlichkeit: Bemerkungen zu ihrer Rolle in der Formierung des Gelehrtenhabitus." In *Der Fehltritt: Vergehen und Versehen un der Vormoderne,* edited by Peter von Moos, 235–50. Cologne: Böhlau, 2001.

———. "Scholars in Households: Refiguring the Learned Habitus, 1480–1550." *Science in Context* 16, no. 1/2 (2003): 9–42.

Ash, Eric H. *Power, Knowledge, and Expertise in Elizabethan England.* Baltimore: Johns Hopkins University Press, 2004.

———. "Queen v. Northumberland, and the Control of Technical Expertise." *History of Science* 39 (2001): 215–40.

Bachmann, Manfred, Harald Marx, and Eberhard Wächtler. *Der Silberne Boden: Kunst und Bergbau in Sachsen.* Stuttgart: Deutsche Verlags-Anstalt; Leipzig: Edition Leipzig, 1990.

Bakewell, P. J. *Mines of Silver and Gold in the Americas.* Brookfield, Vt.: Variorum, 1997.

Baldwin, Martha. "Alchemy and the Society of Jesus in the Seventeenth Century: Strange Bedfellows?" *Isis* 40 (1993): 41–64.

Barrera, Antonio. "Local Herbs, Global Medicines: Commerce, Knowledge, and Commodities in Spanish America." In *Merchants and Marvels: Commerce, Science, and Art in Early Modern Europe,* edited by Pamela H. Smith and Paula Findlen, 163–81. New York: Routledge, 2002.

Bäumel, Jutta. "Electoral Tools and Gardening Implements." In *Princely Splendor: The Dresden Court, 1580–1620,* edited by Dirk Syndram and Antje Scherner, 160–62. Dresden: Staatliche Kunstsammlungen Dresden; Milan: Mondadori Electa, 2004.

Baxandall, Michael. *Painting and Experience in Fifteenth-Century Italy.* Oxford: Oxford University Press, 1972.

Beck, Wolfgang. "Michael Maiers Examen Fucorum Pseudo-Chymicorum: Eine Schrift wider die falschen Alchemisten." PhD diss., Fakultät für Chemie, Biologie und Geowissenschaft der Technischen Universität München, 1992.

Bein, Willy. *Der Stein der Weisen und die Kunst Gold zu machen: Irrtum und Erkenntnis in der Wandlung der Elemente, mitgeteilt nach den Quellen der Vergangenheit und Gegenwart.* Vol. 88, *Voigtländers Quellenbücher.* Leipzig: Voigtländer, ca. 1915.

Belkin, Kristin Lohse. 2005. "Weiditz: (2) Hans Weiditz (ii)." In *Grove Art Online,* http://www.groveart.com/shared/views/article.html?section=art.090980.3 (accessed November 14, 2005, 2005).

Benedict, Philip. "Faith, Fortune and Social Structure in Seventeenth-Century Montpellier." *Past and Present* 152 (1996): 46–78.

Benzenhöfer, Udo. *Johannes de Rupescissa, "Liber de consideratione quintae essentiae omnium rerum" deutsch: Studien zur Alchemia medica des 15. bis 17. Jahrhunderts mit kritischer Edition des Textes.* Stuttgart: Steiner Verlag, 1989.

———. *Paracelsus.* Reinbek bei Hamburg: Rowohlt Taschenbuch Verlag, 1997.

Bernstein, Eckhard. *German Humanism.* Boston: Twayne Publishers, 1983.

Biagioli, Mario. *Galileo Courtier: The Practice of Science in the Culture of Absolutism.* Chicago: University of Chicago Press, 1994.

———. "Galileo's System of Patronage." *History of Science* 28 (1990): 1–62.

Biagioli, Mario, and Peter Galison, eds. *Scientific Authorship: Credit and Intellectual Property in Science.* New York: Routledge, 2002.

Blair, Ann. "Humanist Methods in Natural Philosophy: The Commonplace Book." *Journal of the History of Ideas* 53 (1992): 541–51.

———. "Reading Strategies for Coping with Information Overload, ca. 1550–1700." *Journal of the History of Ideas* 64, no. 1 (2003): 11–28.

Blaschke, Karlheinz. *Sachsen im Zeitalter der Reformation*. Gütersloh: Gütersloher Verlagshaus, 1970.

Boelcke, Willi A. "Das Haus Württemberg und die wirtschaftliche Entwicklung des Landes." In *900 Jahre Haus Württemberg: Leben und Leistung für Land und Volk*, edited by Robert Uhland, 636–77. Stuttgart: W. Kohlhammer, 1985.

Bolton, Henry Carrington. *The Follies of Science at the Court of Rudolph II, 1576–1612*. Milwaukee: Pharmaceutical Review Publishing Co., 1904.

Bönisch, Fritz. *Kursächsische Kartographie bis zum Dreißigjährigen Krieg*. Berlin: Deutscher Verlag der Wissenschaften, 1990.

Braunstein, Phillippe. "Innovations in Mining and Metal Production in Europe in the Late Middle Ages." *Journal of European Economic History* 12, no. 3 (1983): 563–91.

Bullard, Melissa Meriam. "Heroes and Their Workshops: Medici Patronage and the Problem of Shared Agency." *Journal of Medieval and Renaissance Studies* 24 (1994): 179–98.

Buntz, Herwig. "Deutsche alchimistische Traktate des 15. und 16. Jahrhunderts." PhD diss., University of Munich, 1969.

Burckhardt, Daniel. *Albrecht Dürer's Aufenthalt in Basel 1492–1494*. Munich: G. Hirt, 1892.

Christianson, John Robert. *On Tycho's Island: Tycho Brahe and His Assistants (1570–1601)*. Cambridge: Cambridge University Press, 2000.

Cohen, Thomas V., and Elizabeth S. Cohen. *Words and Deeds in Renaissance Rome: Trials before the Papal Magistrates*. Toronto: University of Toronto Press, 1993.

Cramer, M. *M. Johannes Rhenanus, der Pfarrherr und Salzgräfe zu Allendorf*. Halle: Buchhandlung des Waisenhauses, 1879.

Crane, Mark, Richard Raiswell, and Margaret Reeves, eds. *Shell Games: Studies in Scams, Frauds, and Deceits (1300–1500)*. Toronto: Centre for Reformation and Renaissance Studies, 2004.

Crisciani, Chiara. "The Conception of Alchemy as Expressed in the *Pretiosa margarita novella* of Petrus Bonus of Ferrara." *Ambix* 20, no. 3 (1973): 165–81.

Cura, Katrin. "Die Alchemisten und das Gold: Echte und falsche Alchemisten, ihre Laboratorien und Laboranten." *Kultur & Technik* 3 (1998): 34–41.

Czok, Karl, and Reiner Gross. "Das Kurfürstentum, die sächsisch-polnische Union und die Staatsreform (1547–1789)." In *Geschichte Sachsens*, edited by Karl Czok. Weimar: Hermann Böhlaus Nachfolger, 1989.

Darmstaedter, Ernst. *Berg-, Probir- und Kunstbüchlein*. Münchner Beiträge zur Geschichte und Literatur der Naturwissenschaften und Medizin, vols. 2 and 3. Munich: Verlag der Münchner Drucke, 1926.

Daston, Lorraine, and H. Otto Sibum. "Introduction: Scientific Personae and Their Histories." *Science in Context* 16, no. 1/2 (2003): 1–8.

Davis, Natalie Zemon. *Fiction in the Archives.* Stanford, Calif.: Stanford University Press, 1987.

———. *The Return of Martin Guerre.* Cambridge, Mass.: Harvard University Press, 1983.

Debus, Allen G. *The Chemical Philosophy: Paracelsian Science and Medicine in the Sixteenth and Seventeenth Centuries.* New York: Science History Publications, 1977.

DeVun, Leah. "John of Rupescissa and the States of Nature: Science, Apocalypse, and Society in the Late Middle Ages." PhD diss., Columbia University, 2004.

Dobbs, Betty Jo Teeter. *The Foundations of Newton's Alchemy or, "The Hunting of the Green Lyon."* Cambridge: Cambridge University Press, 1975.

———. "From the Secrecy of Alchemy to the Openness of Chemistry." In *Solomon's House Revisited: The Organization and Institutionalization of Science,* edited by Tore Frängsmayr, 75–94. Stockholm: Science History Publications, 1989.

———. *The Janus Faces of Genius: The Role of Alchemy in Newton's Thought.* Cambridge: Cambridge University Press, 1991.

Dopsch, Heinz, Kurt Goldammer, and Peter F. Kramml. *Paracelsus (1493–1541): Keines andern Knecht.* Salzburg: Anton Pustet, 1993.

Dülmen, Richard van. *Theater des Schreckens: Gerichtspraxis und Strafrituale in der frühen Neuzeit.* 3rd ed. Munich: Beck, 1988.

Eamon, William. "Cannibalism and Contagion: Framing Syphilis in Counter-Reformation Italy." *Early Science and Medicine* 1 (February 1998): 1–31.

———. *Science and the Secrets of Nature: Books of Secrets in Medieval and Early Modern Culture.* Princeton, N.J.: Princeton University Press, 1994.

———. "'With the Rules of Life and an Enema': Leonardo Fioravanti's Medical Primitivism." In *Renaissance and Revolution: Humanists, Scholars, Craftsmen and Natural Philosophers in Early Modern Europe,* edited by Frank. V. Field and A. J. L. James, 29–44. Cambridge: Cambridge University Press, 1993.

Ebel, W. "Über das landesherrliche Bergregal." *Zeitschrift für Bergrecht* 109 (1968): 146–83.

Evans, R. J. W. *Rudolf II and His World: A Study in Intellectual History.* Oxford: Oxford University Press, 1973.

Figala, Karin, and Ulrich Neumann. "'Author cui nomen Hermes Malavici': New Light on the Bio-Bibliography of Michael Maier (1569–1622)." In *Alchemy and Chemistry in the 16th and 17th Centuries,* edited by Piyo Rattansi and Antonio Clericuzio, 121–47. Dordrecht: Kluwer Academic, 1994.

Findlen, Paula. "Courting Nature." In *Cultures of Natural History,* edited by Nicholas Jardine, James A. Secord, and E. C. Spary, 57–74. Cambridge: Cambridge University Press, 1996.

———. "Inventing Nature: Commerce, Art, and Science in the Early Modern Cabinet of Wonders." In *Merchants and Marvels: Commerce, Science, and Art in*

Early Modern Europe, edited by Pamela H. Smith and Paula Findlen, 297–323. New York: Routledge, 2002.

———. "Masculine Prerogatives: Gender, Space, and Knowledge in the Early Modern Museum." In *The Architecture of Science,* edited by Peter Galison and Emily Thompson, 29–57. Cambridge, Mass.: MIT Press, 1999.

———. *Possessing Nature: Museums, Collecting, and Scientific Culture in Early Modern Italy.* Berkeley: University of California Press, 1994.

Fleischhauer, Werner. *Renaissance im Herzogtum Württemberg.* Stuttgart: W. Kohlhammer, 1971.

Flynn, Dennis O., and Arturo Giráldez. "Born with a 'Silver Spoon': The Origin of World Trade in 1571." *Journal of World History* 6, no. 2 (1995): 201–21.

Forshaw, Peter. "'Alchemy in the Amphitheatre': Some Considerations of the Alchemical Content of the Engravings in Heinrich Khunrath's *Amphitheatre of Eternal Wisdom* (1609)." In *Art and Alchemy,* edited by Jacob Wamberg, 195–220. Copenhagen: Museum Tusculanum Press, 2006.

Fučíková, Eliška. *Rudolf II and Prague: The Court and the City.* Prague: Prague Castle Administration; London: Thames and Hudson, 1997.

The Fugger News-Letters, Being a Selection of Unpublished Letters from the Correspondents of the House of Fugger during the Years 1568–1605. Edited by Victor Klarwill. Translated by Pauline de Chary. Bodley Head: John Lane, 1928.

Ganzenmüller, W. *Die Alchemie im Mittelalter.* Paderborn, 1938; reprint, Hildesheim, 1967.

Gentilcore, David. "'Charlatans, Mountebanks and Other Similar People': The Regulation and Role of Itinerant Practitioners in Early Modern Italy." *Social History* 20, no. 3 (1995): 297–314.

———. "Contesting Illness in Early Modern Naples: *Miracolati,* Physicians and the Congregation of Rites." *Past and Present* 148 (1995): 117–48.

———. *Healers and Healing in Early Modern Europe.* Manchester: Manchester University Press, 1998.

Geoghegan, D. "Licence of Henry VI to Practise Alchemy." *Ambix* 6 (1957): 10–17.

Ginzburg, Carlo. *The Cheese and the Worms: The Cosmos of a Sixteenth-Century Miller.* Baltimore: Johns Hopkins University Press, 1980.

Glasser, Hannelore. *Artists' Contracts of the Early Renaissance.* Outstanding Dissertations in the Fine Arts. New York: Garland Publishing, 1977.

Goldthwaite, Richard A. *The Building of Renaissance Florence: An Economic and Social History.* Baltimore: Johns Hopkins University Press, 1980.

Grafton, Anthony. *Cardano's Cosmos: The Worlds and Works of a Renaissance Astrologer.* Cambridge, Mass.: Harvard University Press, 1999.

Grimm, Jacob, and Wilhelm Grimm. *Deutsches Wörterbuch.* Leipzig: S. Hirzel, 1854.

Groß, Reiner. *Geschichte Sachsens.* Leipzig: Edition Leipzig, 2001.

Gugenhan, Stefan. *Die landesherrlichen Gärten zu Stuttgart im 16. und 17. Jahrhundert.* Stuttgart: Klett-Cotta, 1997.

Haage, Bernhard Dietrich. *Alchemie im Mittelalter: Ideen und Bilder von Zosimos bis Paracelsus*. Zurich: Artemis & Winkler, 1996.

Halleux, Robert. "L'alchemiste et l'essayeur." In *Die Alchemie in der europäischen Kultur- und Wissenschaftsgeschichte*, edited by Christoph Meinel, 277–92. Wiesbaden: Harrassowitz, 1986.

Hannaway, Owen. *The Chemists and the Word: The Didactic Origins of Chemistry*. Baltimore: Johns Hopkins University Press, 1975.

———. "Laboratory Design and the Aim of Science." *Isis* 77 (1986): 585–610.

Harkness, Deborah E. "Managing an Experimental Household: The Dees of Mortlake and the Practice of Natural Philosophy." *Isis* 88, no. 2 (1997): 247–62.

———. "'Strange' Ideas and 'English' Knowledge: Natural Science Exchange in Elizabethan London." In *Merchants and Marvels: Commerce, Science, and Art in Early Modern Europe*, edited by Pamela H. Smith and Paula Findlen, 137–62. New York: Routledge, 2002.

Heines, Sister Virginia, ed. *Libellus de Alchimia, Ascribed to Albertus Magnus*. Berkeley: University of California Press, 1958.

Henschke, Ekkehard. *Landesherrschaft und Bergbauwirtschaft: Zur Wirtschafts- und Verwaltungsgeschichte des Oberharzer Bergbaugebietes im 15. und 17. Jahrhundert*. Schriften zur Wirtschafts- und Sozialgeschichte, edited by Wolfram Fischer, no. 23. Berlin: Duncker & Humbolt, 1974.

Hill, C. R. "The Iconography of the Laboratory." *Ambix* 22, no. 2 (1975): 102–10.

Hirsch, Rudolf. "The Invention of Printing and the Diffusion of Alchemical and Chemical Knowledge." *Chymia* 3 (1950): 115–42.

Hofacker, Hans-Georg. *". . . sonderlich hohe Künste und vortreffliche Geheimnis": Alchemie am Hof Herzog Friedrichs I. von Württemberg—1593 bis 1608*. Stuttgart: Verein der Freunde des Chemischen Instituts Dr. Flad e. V., 1993.

Holmyard, E. J. *Alchemy*. 2nd ed. Baltimore: Penguin Books, 1968.

Jütte, Robert. *Abbild und soziale Wirklichkeit des Bettler- und Gaunertums zu Beginn der Neuzeit: Sozial-, mentalitäts- und sprachgeschichtliche Studien zum Liber vagatorum (1510)*. Cologne: Böhlau, 1988.

———. *Poverty and Deviance in Early Modern Europe*. Cambridge: Cambridge University Press, 1994.

Karant-Nunn, Susan C. "The Women of the Saxon Silver Mines." In *Women in Reformation and Counter-Reformation Europe*, edited by Sherrin Marshall, 29–46. Bloomington: Indiana University Press, 1989.

Karpenko, Vladimír. "Alchemy as 'Donum Dei.'" *Hyle: An International Journal for the Philosophy of Chemistry* 4 (1998): 63–80.

———. "Bohemian Nobility and Alchemy in the Second Half of the 16th Century: Wilhelm of Rosenberg and Two Alchemists." *Cauda Pavonis: The Hermetic Text Society Newsletter* 15, no. 2 (1996).

———. "The Chemistry and Metallurgy of Transmutation." *Ambix* 39, no. 2 (1992): 47–62.

Kassell, Lauren. "Reading for the Philosophers' Stone." In *Books and the Sciences in History*, edited by Marina Frasca-Spada and Nick Jardine, 132–50. Cambridge: Cambridge University Press, 2000.

Kaufman, Thomas DaCosta. *The Mastery of Nature: Aspects of Art, Science, and Humanism in the Renaissance*. Princeton, N.J.: Princeton University Press, 1993.

Kellenbenz, Hermann. *Deutsche Wirtschaftsgeschichte*. 2 vols. Munich: Beck, 1977.

———. *The Rise of the European Economy: An Economic History of Continental Europe from the Fifteenth to the Eighteenth Century*. New York: Holmes and Meier Publishers, 1976.

Koch, Hugo. *Sächsische Gartenkunst*. 2nd ed. Beucha: Sax-Verlag, ca. 1999.

Könneker, Barbara. *Satire im 16. Jahrhundert: Epoche, Werke, Wirkung*. Munich: C. H. Beck, 1991.

Kopp, Hermann. *Die Alchemie in älterer und neuerer Zeit*. 2 vols. Heidelberg: Carl Winter's Universitätsbuchhandlung, 1886.

Kötzschke, Rudolf, and Hellmut Kretzschmar. *Sächsische Geschichte*. Frankfurt am Main: Verlag Wolfgang Weidlich, 1965.

Kraschewski, Hans-Joachim. "Heinrich Cramer von Clausbruch und seine Handelsverbindungen mit Herzog Julius von Braunschweig-Wolfenbüttel: Zur Geschichte des Fernhandels mit Blei und Vitriol in der zweiten Hälfte des 16. Jahrhunderts." *Braunschweigisches Jahrbuch* 66 (1985): 115–28.

———. "Der 'ökonomische' Fürst: Herzog Julius als Unternehmer-Verleger der Wirtschaft seines Landes, besonders des Harz-Bergbaus." In *Staatsklugheit und Frömmigkeit: Herzog Julius von Braunschweig-Lüneburg—ein norddeutscher Landesherr des 16. Jahrhunderts*, edited by Christa Graefe and the Herzog August Bibliothek (Wolfenbüttel), 41–58. Weinheim: VCH Verlagsgesellschaft, Acta Humaniora, 1989.

———. *Wirtschaftspolitik im deutschen Territorialstaat des 16. Jahrhunderts: Herzog Julius von Braunschweig-Wolfenbüttel*. Edited by Ingomar Bog. Neue Wirtschaftsgeschichte, vol. 15. Cologne: Böhlau Verlag, 1978.

Kubátová, Ludmila, Hans Prescher, and Werner Weisbach. *Lazarus Ercker (1528/30–1594): Probierer, Berg- und Münzmeister in Sachsen, Braunschweig und Böhmen*. Leipzig: Deutscher Verlag für Grundstoffindustrie, 1994.

Laube, A. *Studien über den erzgebirgischen Silberbergbau von 1470 bis 1546*. Berlin: Akademie-Verlag, 1974.

Lehrich, Christopher I. *The Language of Demons and Angels: Cornelius Agrippa's Occult Philosophy*. Leiden: Brill, 2003.

Linden, Stanton J. *Darke Hierogliphicks: Alchemy in English Literature from Chaucer to the Restoration*. Lexington: University Press of Kentucky, 1996.

Loats, Carol L. "Gender and Work in Paris: The Evidence of Employment Contracts, 1540–1560." *Proceedings of the Western Society for French History* 20 (1993): 25–37.

Lohmeier, Dieter. "Heinrich Rantzau und die Adelskultur der frühen Neuzeit."

In *Arte et Marte: Studien zur Adelskultur des Barockzeitalters in Schweden, Dänemark und Schleswig-Holstein,* edited by D. Lohmeier, 67–84. Neumünster: K. Wachholtz, 1978.

Long, Pamela O. "The Openness of Knowledge: An Ideal and Its Context in 16th-Century Writings on Mining and Metallurgy." *Technology and Culture* 32, no. 2 (1991): 318–55.

———. *Openness, Secrecy, Authorship: Technical Arts and the Culture of Knowledge from Antiquity to the Renaissance.* Baltimore: Johns Hopkins University Press, 2001.

Luhrmann, T. M. *Persuasions of the Witch's Craft: Ritual Magic in Contemporary England.* Cambridge, Mass.: Harvard University Press, 1989.

Martinón-Torres, Marcos, and Thilo Rehren. "Alchemy, Chemistry and Metallurgy in Renaissance Europe: A Wider Context for Fire-Assay Remains." *Historical Metallurgy* 39, no. 1 (2005): 14–28.

Matton, Sylvain. "L'influence de l'humanisme sur la tradition alchimique." *Micrologus* 3 (1995): 279–345.

Mauss, Marcel. "Une catégorie de l'esprit humain: La notion de personne, celle de 'moi': Un plan de travail." *Journal of the Royal Anthropological Institute* 68 (1938): 236–81.

Molenda, Danuta. "Technological Innovation in Central Europe between the XIVth and the XVIIth Centuries." *Journal of European Economic History* 17, no. 1 (1988): 63–84.

Moran, Bruce T. *The Alchemical World of the German Court: Occult Philosophy and Chemical Medicine in the Circle of Moritz of Hessen (1572–1632).* Stuttgart: Franz Steiner Verlag, 1991.

———. "The Alchemist's Reality: Problems and Perceptions of a German Alchemist in the 17th Century." *Halcyon: A Journal of the Humanities* 9 (1987): 133–48.

———. *Distilling Knowledge: Alchemy, Chemistry, and the Scientific Revolution.* Cambridge, Mass.: Harvard University Press, 2005.

———. "German Prince-Practitioners: Aspects in the Development of Courtly Science." *Technology and Culture* 22, no. 2 (1981): 253–74.

———, ed. *Patronage and Institutions: Science, Technology, and Medicine at the European Court, 1500–1750.* Rochester, N.Y.: Boydell Press, 1991.

Morys, Peter. "Leonhard Thurneissers *De transmutatione veneris in solem.*" In *Die Alchemie in der europäischen Kultur- und Wissenschaftsgeschichte,* edited by Christoph Meinel, 85–98. Wiesbaden: Otto Harrassowitz, 1986.

———. *Medizin und Pharmazie in der Kosmologie Leonhard Thurneissers zum Thurm (1531–1596).* Edited by Rolf Winau and Heinz Müller-Dietz. Abhandlungen zur Geschichte der Medizin und der Naturwissenschaften, vol. 43. Husum: Matthiesen, 1982.

Moss, Ann. *Printed Commonplace-Books and the Structuring of Renaissance Thought.* Oxford: Clarendon Press, 1996.

Muir, Edward, and Guido Ruggiero. *History from Crime*. Baltimore: Johns Hopkins University Press, 1994.

Munro, John H. "The Central European Silver Mining Boom, Mint Outputs, and Prices in the Low Countries and England, 1450–1550." In *Money, Coins, and Commerce: Essays in the Monetary History of Asia and Europe from Antiquity to Modern Times,* edited by Eddy Van Cauwenberghe, 119–83. Leuven: Leuven University Press, 1991.

———. "Patterns of Trade, Money, and Credit." In *Handbook of European History, 1400–1600: Late Middle Ages, Renaissance, and Reformation,* edited by Thomas A. Brady, Heiko Augustinus Oberman, and James D. Tracy, 147–95. Leiden: E. J. Brill, 1994.

Newman, William R. "Alchemical Symbolism and Concealment: The Chemical House of Libavius." In *The Architecture of Science,* edited by Peter Galison and Emily Thompson, 59–78. Cambridge, Mass.: MIT Press, 1999.

———. "Alchemy, Assaying, and Experiment." In *Instruments and Experimentation in the History of Chemistry,* edited by Frederic L. Holmes and Trevor Levere, 35–54. Cambridge, Mass.: MIT Press, 2000.

———. *Atoms and Alchemy: Chymistry and the Experimental Origins of the Scientific Revolution.* Chicago: University of Chicago Press, 2006.

———. *Gehennical Fire: The Lives of George Starkey, an American Alchemist in the Scientific Revolution.* Cambridge, Mass.: Harvard University Press, 1994.

———. "The Philosopher's Egg: Theory and Practice in the Alchemy of Roger Bacon." *Micrologus* 3 (1995): 75–101.

———. "The Place of Alchemy in the Current Literature on Experiment." In *Experimental Essays: Versuche zum Experiment,* edited by Michael Heidelberger and Friedrich Steinle, 9–33. Baden-Baden: Nomos, 1998.

———. *Promethean Ambitions: Alchemy and the Quest to Perfect Nature.* Chicago: University of Chicago Press, 2004.

———, ed. *The Summa Perfectionis of Pseudo-Geber: A Critical Edition, Translation and Study.* Leiden: E. J. Brill, 1991.

———. "Technology and the Alchemical Debate in the Late Middle Ages." *Isis* 80 (1989): 430–37.

Newman, William R., and Lawrence M. Principe. *Alchemy Tried in the Fire: Starkey, Boyle, and the Fate of Helmontian Chymistry.* Chicago: University of Chicago Press, 2002.

———. "Alchemy vs. Chemistry: The Etymological Origins of a Historiographic Mistake." *Early Science and Medicine* 3 (1998): 32–65.

Nick, Friedrich. *Stuttgarter Chronik und Sagenbuch: Eine Sammlung denkwürdiger Begebenheiten, Geschichten und Sagen der Stadt Stuttgart und ihrer Gemarkung.* Stuttgart: Emil Gutzkow, Verlagsbuchhandlung, 1875.

Nummedal, Tara E. "Alchemical Reproduction and the Career of Anna Maria Zieglerin." *Ambix* 48 (2001): 56–68.

Nutton, Vivian. *Medicine at the Courts of Europe, 1500–1837*. Wellcome Institute Series in the History of Medicine. London; New York: Routledge, 1990.

Obrist, Barbara. "Die Alchemie in der mittelalterlichen Gesellschaft." In *Die Alchemie in der europäishen Kultur- und Wissenschaftsgeschichte,* edited by Christoph Meinel, Wolfenbütteler Forschungen, vol. 32. Wiesbaden: Otto Harrassowitz, 1986.

Ogrinc, Will H. L. "Western Society and Alchemy from 1200–1500." *Journal of Medieval History* 6 (1980): 103–37.

Oldridge, Darren. *Strange Histories: The Trial of the Pig, the Walking Dead, and Other Matters of Fact from the Medieval and Renaissance Worlds*. London: Routledge, 2005.

Pagel, Walter. *Paracelsus: An Introduction to Philosophical Medicine in the Era of the Renaissance*. 2nd ed. Basel: Karger, 1982.

Partington, J. R. "Albertus Magnus on Alchemy." *Ambix* 1, no. 1 (1937): 3–20.

Pereira, Michela. "Alchemy and the Use of Vernacular Languages in the Late Middle Ages." *Speculum* 74, no. 2 (1999): 336–56.

———. "*Mater Medicinarum:* English Physicians and the Alchemical Elixir in the Fifteenth Century." In *Medicine from the Black Death to the French Disease,* edited by Roger French, 26–52. Aldershot: Ashgate, 1998.

———. "Teorie dell'elixir nell'alchemia medievale." *Micrologus* 3 (1995): 103–48.

Pfeilsticker, Walter. *Neues Württembergisches Dienerbuch*. Stuttgart: J. G. Cotta'sche Buchhandlung, 1957–63.

Piasecki, Peter. *Das deutsche Salinenwesen, 1550–1650: Invention—Innovation—Diffusion*. Idstein: Schulz-Kirchner Verlag, 1987.

Pomata, Gianna. *Contracting a Cure: Patients, Healers, and the Law in Early Modern Bologna*. Baltimore: Johns Hopkins University Press, 1998.

Priesner, Claus, and Karin Figala, eds. *Alchemie: Lexikon einer hermetischen Wissenschaft*. Munich: Beck, 1998.

Principe, Lawrence M. *The Aspiring Adept: Robert Boyle and His Alchemical Quest*. Princeton, N.J.: Princeton University Press, 1998.

———. "Robert Boyle's Alchemical Secrecy: Codes, Ciphers and Concealments." *Ambix* 39 (1992): 63–74.

Principe, Lawrence M., and William R. Newman. "Some Problems with the Historiography of Alchemy." In *Secrets of Nature: Astrology and Alchemy in Early Modern Europe,* edited by William R. Newman and Anthony Grafton. Cambridge, Mass.: MIT Press, 2001.

Pumphrey, Stephen, and Frances Dawbarn. "Science and Patronage in England, 1570–1625." *History of Science* 42, no. 2, no. 136 (2004): 137–87.

Rankin, Alisha. "Becoming an Expert Practitioner: Court Experimentalism and the Medical Skills of Anna of Saxony (1532–85)." *Isis* 98 (2007): 23–53.

———. "Medicine for the Uncommon Woman: Experience, Experiment, and Exchange in Early Modern Germany." PhD diss., Harvard University, 2005.

Raupp, Hans-Joachim. "Die Illustrationen zu Francesco Petrarca, 'Von der Artzney Bayder Glueck des Guten und Widerwertigen' (Augsburg 1532)." *Wallraf-Richartz-Jahrbuch (Westdeutsches Jahrbuch für Kunstgeschichte)* 45 (1984): 59–112.

Rawski, Conrad H., ed. *Remedies for Fortune Fair and Foul: A Modern English Translation of De remediis utriusque fortune, with a Commentary.* Bloomington: Indiana University Press, 1991.

Redlich, Fritz. "Der deutsche fürstliche Unternehmer, eine typische Erscheinung des 16. Jahrhunderts." *Tradition: Zeitschrift für Firmengeschichte und Unternehmer-Biographie* 3, no. 1 (1958): 17–23, 98–112.

Rhamm, Albert. *Die betrüglichen Goldmacher am Hofe des Herzogs Julius von Braunschweig: Nach den Processakten.* Wolfenbüttel: Julius Zwißler, 1883.

Roberts-Jones, Philippe, and Françoise Roberts-Jones. *Pieter Bruegel.* New York: Harry N. Abrams, 2002.

Ruggiero, Guido. *Binding Passions: Tales of Magic, Marriage, and Power at the End of the Renaissance.* Oxford: Oxford University Press, 1993.

———. "The Strange Death of Margarita Marcellini: *Male,* Signs, and the Everyday World of Pre-modern Medicine." *American Historical Review* 106, no. 4 (2001): 1141–58.

Sabean, David. *Power in the Blood: Popular Culture and Village Discourse in Early Modern Germany.* Cambridge: Cambridge University Press, 1984.

Sandman, Alison. "Mirroring the World: Sea Charts, Navigation, and Territorial Claims in Sixteenth-Century Spain." In *Merchants and Marvels: Commerce, Science, and Art in Early Modern Europe,* edited by Pamela H. Smith and Paula Findlen, 83–108. New York: Routledge, 2002.

Sauer, Paul. *Herzog Friedrich I. von Württemberg 1557–1608: Ungestümer Reformer und weltgewandter Autokrat.* Munich: Deutsche Verlags-Anstalt, 2003.

Scheidig, Walther. *Die Holzschnitte des Petrarca-Meisters: Zu Petrarcas Werk von der Artzney bayder Glück des guten und widerwärtigen, Augsburg 1532.* Berlin: Henschelverlag/Deutsche Akademie der Künste, 1955.

Schott, Heinz, and Ilana Zinguer, eds. *Paracelsus und seine internationale Rezeption in der frühen Neuzeit: Beiträge zur Geschichte des Paracelsismus.* Brill's Studies in Intellectual History, vol. 86. Leiden: Brill, 1998.

Schramm, Petra. *Die Alchemisten: Gelehrte—Goldmacher—Gaukler: Ein dokumentarischer Bildband.* Wiesbaden: Edition Rarissima, 1984.

Shackelford, Jole. *A Philosophical Path for Paracelsian Medicine: The Ideas, Intellectual Context, and Influence of Petrus Severinus, 1540–1602.* Copenhagen: Museum Tusculanum, 2004.

———. "Tycho Brahe, Laboratory Design, and the Aim of Science: Reading Plans in Context." *Isis* 84 (1993): 211–30.

Shapin, Steven. "The House of Experiment in Seventeenth-Century England." *Isis* 79 (1988): 373–404.

———. *A Social History of Truth: Civility and Science in Seventeenth-Century*

England. Science and Its Conceptual Foundations. Chicago: University of Chicago Press, 1994.

Shapin, Steven, and Simon Schaffer. *Leviathan and the Air-Pump: Hobbes, Boyle, and the Experimental Life*. Princeton, N.J.: Princeton University Press, 1985.

Smith, Pamela H. "Alchemy as a Language of Mediation at the Habsburg Court." *Isis* 85 (March 1994): 1–25.

———. *The Body of the Artisan: Art and Experience in the Scientific Revolution*. Chicago: University of Chicago Press, 2004.

———. *The Business of Alchemy: Science and Culture in the Holy Roman Empire*. Princeton, N.J.: Princeton University Press, 1994.

———. "Consumption and Credit: The Place of Alchemy in Johann Joachim Becher's Political Economy." In *Alchemy Revisited*, edited by Z. R. W. M. von Martels, 215–21. Leiden: Brill, 1990.

———. "Curing the Body Politic: Chemistry and Commerce at Court, 1664–70." In *Patronage and Institutions: Science, Technology, and Medicine at the European Court, 1500–1750*, edited by Bruce T. Moran, 195–209. Rochester, N.Y.: Boydell Press, 1991.

———. "Laboratories." In *The Cambridge History of Science*, vol. 3, *Early Modern Europe*, edited by Lorraine Daston and Katharine Park, 289–305. Cambridge: Cambridge University Press, 2006.

Soukup, R. W., S. von Osten, and H. Mayer. "Alembics, Cucurbits, Phials, Crucibles: A 16th-Century Docimastic Laboratory Excavated in Austria." *Ambix* 40, no. 1 (1993): 25.

Soukup, Rudolf Werner, and Helmut Mayer. *Alchemistisches Gold, paracelsistische Pharmaka*. Edited by Helmuth Grössing, Karl Kadletz, and Marianne Klemun. Perspektiven der Wissenschaftsgeschichte, vol. 10. Vienna: Böhlau Verlag, 1997.

Spierenburg, Pieter. "The Body and the State: Early Modern Europe." In *The Oxford History of the Prison: The Practice of Punishment in Western Society*, edited by Norval Morris and David J. Rothman. New York: Oxford University Press, 1995.

Spitzer, Gabriele. *". . . und die Spree führt Gold": Leonhard Thurneysser zum Thurn, Astrologe—Alchimist—Arzt und Drucker im Berlin des 16. Jahrhunderts*. Beiträge aus der Staatsbibliothek zu Berlin, Pruessischer Kulturbesitz, Band 3. Wiesbaden: Reichert, 1996.

Spooner, F. C. "The Economy of Europe, 1559–1609." In *The New Cambridge Modern History, Vol. III*, edited by R. B. Wernham, 14–43. Cambridge: Cambridge University Press, 1968.

Starenko, Peter Elsel. "In Luther's Wake: Duke John Frederick II of Saxony, Angelic Prophecy, and the Gotha Rebellion of 1567." PhD thesis, University of California, Berkeley, 2002.

Strieder, Jacob. *Jacob Fugger the Rich, Merchant and Banker of Augsburg, 1459–1525*. New York: Adelphi Co., 1931.

———. *Studien zur Geschichte kapitalistischer Organisationsformen: Monopole, Kartelle und Aktiengesellschaften im Mittelalter und zu Beginn der Neuzeit.* 2nd expanded ed. New York: B. Franklin, 1971.

Striedinger, Ivo. *Der Goldmacher Marco Bragadino.* Munich: Theodor Ackerman, 1928.

Sudhoff, Karl. *Paracelsus: Ein deutsches Lebensbild aus den Tagen der Renaissance.* Leipzig: Bibliographisches Institut, 1936.

Suhling, Lothar. "'Philosophisches' in der frühneuzeitlichen Berg- und Hüttenkunde: Metallogenese und Transmutation aus der Sicht montanischen Erfahrungswissens." In *Die Alchemie in der europäischen Kultur- und Wissenschaftsgeschichte,* edited by Christoph Meinel, 293–313. Wiesbaden: Harrassowitz, 1986.

Svatek, Josef. *Culturhistorische Bilder aus Böhmen.* 2 vols. Vienna: Wilhelm Braunmüller, k. k. Hof- und Universitätsbuchhändler, 1879.

Syndram, Dirk, and Antje Scherner, eds. *Princely Splendor: The Dresden Court, 1580–1620.* Dresden: Staatliche Kunstsammlungen Dresden; Milan: Mondadori Electa, 2004.

Taylor, F. Sherwood. *The Alchemists, Founders of Modern Chemistry.* New York: H. Schuman, 1949.

Telle, Joachim. "Der Alchemist im Rosengarten: Ein Gedicht von Christoph von Hirschenberg für Landgraf Wilhelm IV. von Hessen-Kassel und Graf Wilhlem von Zimmern." *Euphorion* 71 (1977): 283–305.

———, ed. *Analecta Paracelsica: Studien zum Nachleben Theophrast von Hohenheims im deutschen Kulturgebiet der frühen Neuzeit.* Heidelberger Studien zur Naturkunde der frühen Neuzeit, vol. 4. Stuttgart: F. Steiner, 1994.

———. "Johannes Huser und der Paracelsismus im 16. Jahrhundert." In *Paracelsus (1493–1541): Keines andern Knecht,* edited by Heinz Dopsch, Kurt Goldammer, and Peter F. Kramml. Salzburg: Anton Pustet, 1993.

———, ed. *Rosarium philosophorum: Ein alchemisches Florilegium des Spätmittelalters, Faksimile der illustrierten Erstausgabe Frankfurt 1550.* Translated by Lutz Claren and Joachim Huber. 2 vols. Weinheim: VCH Verlagsgesellschaft, 1992.

———, ed. *Parerga Paracelsica: Paracelsus in Vergangenheit und Gegenwart.* Heidelberger Studien zur Naturkunde der frühen Neuzeit, vol. 3. Stuttgart: Franz Steiner Verlag, 1991.

———. *Sol und Luna: Literar- und alchemiegeschichtliche Studien zu einem altdeutschen Bildgedicht.* Hürtgenwald: Guido Pressler Verlag, 1980.

———. "'Vom Stein der Weisen': Ein alchemoparacelsstische Lehrdichtung des 16. Jahrhunderts." In *Analecta Paracelsica: Studien zum Nachleben Theophrast von Hohenheims im deutschen Kulturgebiet der frühen Neuzeit,* edited by Joachim Telle, 167–212. Stuttgart: F. Steiner, 1994.

Thorndike, Lynn. "Alchemy during the First Half of the Sixteenth Century." *Ambix* 2 (1938): 26–37.

———. *A History of Magic and Experimental Science.* 8 vols. New York: Columbia University Press, 1934–58.

Tilton, Hereward. *The Quest for the Phoenix: Spiritual Alchemy and Rosicrucianism in the Work of Count Michael Maier (1569–1622).* Edited by Christoph Markschies and Gerhard Mueller. Arbeiten zur Kirchengeschichte 88. Berlin: Walter de Gruyter, 2003.

Töllner, Ralf. *Der unendliche Kommentar: Untersuchungen zu vier ausgewählten Kupferstichen aus Heinrich Khunraths "Amphitheatrum sapientiae aeternae solius verae" (Hanau 1609).* Ammersbek bei Hamburg: Verlag an der Lottbek, 1991.

Treue, Wilhelm. "Das Verhältnis von Fürst, Staat und Unternehmer in der Zeit des Merkantilismus." *Vierteljahrschrift für Sozial- und Wirtschaftsgeschichte* 44 (1957): 26–56.

Trunz, Erich. *Wissenschaft und Kunst im Kreise Kaiser Rudolfs II. 1576–1612.* Neumünster: Karl Wachholtz Verlag, 1992.

Van Cleve, John Walter. *The Problem of Wealth in the Literature of Luther's Germany.* Columbia, S.C.: Camden House, 1991.

Van Lennep, Jacques. "An Alchemical Message in Brueghel's Prints?" In *The Prints of Pieter Brueghel the Elder,* edited by David Freedberg, 66–79. Tokyo: Tokyo Shibum, 1989.

Wagner, Theodor. "Wissenschaftlicher Schwindel aus dem südlichen Böhmen." *Mittheilungen des Vereins für Geschichte der Deutschen in Böhmen* 16 (1878): 112–23.

Watanabe-O'Kelly, Helen. *Court Culture in Dresden: From Renaissance to Baroque.* Houndmills: Palgrave, 2002.

Weber, Karl von. *Aus vier Jahrhunderten: Mittheilungen aus dem Haupt-Staatsarchive zu Dresden.* Leipzig: Verlag von Bernhard Tauchnitz, 1858.

Weyer, Jost. "Der Alchemist im lateinischen Mittelalter (13. bis 15. Jahrhundert)." In *Der Chemiker im Wandel der Zeiten: Skizzen zur geschichtlichen Entwicklung des Berufbildes,* edited by Eberhard Schmauderer and Gesellschaft Deutscher Chemiker, Fachgruppe Geschichte der Chemie, 11–41. Weinheim: Verlag Chemie, 1973.

———. "Der 'Goldmacher' Michael Polhaimer: Alchemistischer Betrüger am Hof des Grafen Wolfengang II. von Hohenlohe." *Beitrage zur Landeskunde, Regelmäßige Beilage zum Staatsanzeiger für Baden-Württemberg* 4 (1993): 7–11.

———. *Graf Wolfgang II. von Hohenlohe und die Alchemie: Alchemistische Studien in Schloß Weikersheim, 1587–1610.* Edited by Historischen Verein für Württembergisch Franken, Stadtarchiv Schwäbisch Hall and Hohenlohe-Zentralarchiv Neuenstein. Forschungen aus Württembergisch Franken, vol. 39. Sigmaringen: Jan Thorbecke Verlag, 1992.

Winkler, Friedrich. *Dürer und die Illustrationen zum Narrenschiff: Die Baseler und Straßburger Arbeiten des Künstlers und der altdeutsche Holzschnitt.* Berlin: Deutscher Verein für Kunstwissenschaft, 1951.

Wr., Prof. H. "Die Goldmacherbande am Hofe des Herzogs Julius von Braunschweig in Wolfenbüttel." *Niedersachsen* 14 (1908/1909): 346–51.

Wright, William J. "The Nature of Early Capitalism." In *Germany: A New Social and Economic History,* edited by Bob Scribner, 181–208. London: Arnold, 1996.

Yates, Frances A. *Giordano Bruno and the Hermetic Tradition.* Chicago: University of Chicago Press, 1964.

———. *The Rosicrucian Enlightenment.* London: Routledge and Kegan Paul, 1972.

Zachar, Otakar. *O alchymii a českých alchymistech.* Prague: V. Kotrba, 1911.

———. "Z dějin alchymie v Čechách. I. Bavor mladši Rodovský z Hustiřan, alchymista český." *Časopis Musea Kralovstí Českého* XXI-II (1899–1900): 157–63.

INDEX